TECHNOLOGY SERIES   Technical Memorandum No. 10

Small-scale processing of beef

Prepared under the joint auspices of the International Labour Office and the United Nations Environment Programme

International Labour Office   Geneva

ISBN 92-2-105050-5
ISSN 0252-2004

First published 1985

Printed in Switzerland

PREFACE

This technical memorandum is the fifth in a series of publications dealing with food processing technologies[1]. The object of the series is to acquaint small-scale producers with alternative production techniques for specific products and processes with a view to helping them to choose and apply those techniques which are most appropriate to local socio-economic conditions.

A large number of developing countries raise cattle but continue to import large amounts of processed beef (i.e. corned beef). It is estimated that products worth over US$170 million are imported by developing countries each year, mostly from industrialised countries. Yet developing countries are perfectly capable of processing their own beef, thus saving scarce foreign exchange and generating much needed new employment opportunities. Furthermore, the local manufacture of beef products should better take into consideration local tastes and customs and the purchasing power of potential consumers of these products.

There are many reasons which may explain the large imports of processed beef products by developing countries and the limited production of these products by local meat processors. Interested readers will find a brief analysis of these reasons in Chapter I. One such reason seems to be the lack of information on small-scale beef processing technologies. It is hoped that this memorandum will help bridge this information gap and will induce the local processing of beef in small-scale plants or as part of a butchery business.

---

[1] Published memoranda on food processing cover the following products: oil extraction from groundnuts and copra, small-scale processing of fish, small-scale maize milling and small-scale processing of pork. Three other memoranda are at various stages of preparation: small-scale fruit processing, small-scale vegetable processing and small-scale grain storage.

This memorandum covers the production of seven types of standard beef products: biltong, charqui, fresh beefburgers, various types of cooked and cured beef, chili con carne, frankfurters and beef cervelat. Recipes and processing techniques are described in detail for the above types of beef products. These may be easily adapted for the production of other local specialties not covered by this memorandum (e.g. through the addition of other ingredients or small variations in the various processing stages). Thus, with few exceptions, this memorandum should be of interest to the majority of beef processors in developing countries.

The scale of production described in this memorandum does not exceed 2 tonnes of product per week. Production units with an output equal to or lower than 2 tonnes per week are defined as small-scale units. This is an arbitrary definition which does not take into consideration special circumstances under which beef processing is carried out in various countries. Thus, units producing 2 tonnes of processed beef per week may be considered medium-scale units in some countries or "micro units" in other countries. However, from a purely technological point of view (i.e taking into consideration available meat processing equipment), the upper limit for small-scale meat processing units is approximately 2 tonnes of product per week.

The processing techniques described in the memorandum are fairly simple and do not require automated or highly sophisticated equipment. However, most pieces of equipment are equipped with a small electric motor in view of the very low productivity of manually operated equipment. Unlike the other technical memoranda, no list of equipment suppliers is provided in this memorandum. The reason for this is that there is a very large number of meat processing equipment suppliers which are represented in most developing countries. Readers who may face difficulties locating equipment suppliers can obtain names and addresses from one of the directories or journals listed at the end of the memorandum.

Beef processing may be operated as an entirely separate and independent business or may be an integral part of a slaughterhouse complex or of a butchery business. In general, circumstances (e.g. volume of demand for beef products, location of production) will favour one or the other organisation of production. However, this technical memorandum focuses on the establishment of beef processing units operating as separate businesses. This choice is made in order to facilitate the exposition of the various technical factors and requirements of beef processing and to allow for a separate assessment of the viability of a projected plant. However, the information provided should allow, with some adaptations, the assessment of the technical and economic feasibility of processing units attached to a slaughterhouse or butchery business.

Two chapters of this memorandum (Chapters I and VI) are of particular interest to public planners, project evaluators from industrial development agencies and financial institutions. Chapter I analyses demand and supply of processed beef in developing countries, including imports of beef products and reasons for the limited supply of locally processed meat. Chapter VI evaluates the effects of alternative meat processing technologies on employment generation, foreign exchange savings, capital expenditures and rural industrialisation. A special section of this chapter is also devoted to the effects of alternative technologies on the environment, especially in relation to the disposal of wastes generated by small-scale and large-scale plants. These two chapters should help public planners to formulate policies and measures in favour of appropriate scales of production and meat processing technologies.

The remaining chapters (Chapters II to V) are of particular interest to small-scale meat processors. Chapter II describes the main raw materials (meat cuts, casings, spices, etc.) used in the manufacture of beef products, while Chapter III describes the various pieces of equipment used in meat processing and provides some guide-lines for the design of a meat processing unit, including a suggested plant layout. Chapter IV describes in detail the techniques used for the production of the selected beef products, including the recipe for a given batch of products and a description of the various processing stages. Finally, Chapter V suggests a methodological framework for the evaluation of alternative scales of production and processing technologies. This framework is applied to two production models for illustrative purposes.

A questionnaire is attached at the end of the memorandum for those readers who may wish to send to the ILO their comments and suggestions on the content and usefulness of this publication. These will be taken into consideration in the preparation of future technical memoranda.

This technical memorandum was prepared by the Tropical Development and Research Institute (London) in collaboration with Mr. M. Allal, staff member in charge of the Technology Series within the Technology and Employment Branch of the ILO.

A. S. Bhalla,

Chief,

Technology and Employment Branch.

CONTENTS

APPENDICES

A C K N O W L E D G E M E N T S

The publication of this technical memorandum has been made possible by a grant from the United Nations Environment Programme. The International Labour Office gratefully acknowledges this generous support.

Figures III.2, III.3, III.4, III.5, III.6, III.9: (c) British Crown Copyright, courtesy of TDRI.

CHAPTER I

SMALL-SCALE BEEF PROCESSING IN DEVELOPING
COUNTRIES: SOME GENERAL CONSIDERATIONS

The purpose of this chapter is briefly to analyse demand, production and trade of processed beef in developing countries, to identify the various types of organisation of production in the beef processing sector, and to indicate the scales of production, processing techniques and beef products covered by this memorandum.

I. DEMAND, TRADE AND PRODUCTION

I.1 Imports of processed beef by developing countries

Comprehensive information is not available on production and consumption of traditional meat products in developing countries. However, trade statistics indicate that steadily expanding markets for many kinds of meat products exist in these countries. Total imports of processed meat by all developing regions are substantial and rapidly growing (see table I.1). They averaged 131,000 tonnes per annum for the period 1965-69 and 170,000 tonnes per annum between 1970 and 1974, and exceeded 200,000 tonnes per annum by 1977. Total value of processed meat imports increased from an average of US$106 million in 1965-69 to US$401 million in 1978, although an estimated 80 per cent of this increase is attributable to inflation.

Apart from bacon and 'bone-in' ham, it is difficult to ascertain the animal derivation of the imports shown in the table. However, estimates for the second half of the 1970s indicate that about 43 per cent (by value)

Table I.1

Imports of meat products by developing countries, 1965-78

(Average per annum)

Product	SITC (R2)	1965-69	1970-74	1975	1976	1977	1978
		(in thousands of tonnes)					
Canned or prepared meat	014.9	84	118	139	136	153	149
Bacon and ham	012.1	24	25	24	24	22	24
Sausages	014.2	14	17	17	17	20	22
Other dried, salted, smoked meat	012.9	9	10	12	16	14	12
Total		131	170	192	193	209	207
Total value (in millions of US$, mainly c.i.f., 1984 prices)		106	183	295	311	370	401

Source: FAO Trade Yearbook (see Bibliography)

of total developing country imports of meat products are based on beef and 53 per cent on pig-meat, although these percentages may vary appreciably from year to year. The remainder (about 5 per cent) are based mainly on meat from sheep, goat, poultry and game. Thus, developing countries could have saved, in 1978, the equivalent of US$172 million in scarce foreign exchange had they processed their own beef. Since most of these imports came from developed countries, the above gross savings would not have been made at the expense of other developing countries. In fact, actual net saving in foreign exchange to developing countries would have been lower than US$172 million (in 1978) because additional imports, mostly from developed countries (e.g. processing machinery, packaging materials, animal feed), would have been necessary for the local processing of beef. However, even though an accurate estimate cannot be made, significant net savings in foreign exchange could probably be achieved by some developing countries if processed beef is produced locally instead of being imported.

I.2 Processing of beef in developing countries

Despite the fact that some developing countries raise considerable numbers of cattle, meat processing plants have not developed on a large scale. A few exceptions to this situation mainly relate to large-scale canning operations. Major developments have been confined to the processing of beef in some countries of South America and, to a lesser extent, of Africa. In the past, plants tended to be owned by foreign firms and production was primarily for export. While this situation is changing in favour of local ownership, most beef is still produced and consumed locally in developing countries, either fresh or as distinctive products of traditional preservation processes (e.g. biltong, charqui).

However, the considerable volume of imports of processed beef by developing countries suggest that the latter could considerably expand production in order to satisfy local demand. What may, therefore, explain the lack of response from local meat processors? One explanation could be the characteristics of imported beef products (e.g. beefburgers). In general, these require high quality raw materials and are much more difficult to manufacture, store and distribute than traditional products since they are more susceptible to bacterial spoilage during processing and storage. An accurate control of cooking temperatures and times as well as the maintenance of an appropriate storage temperature are critical. High standards of hygiene at all stages of production and distribution are also essential. Thus, potential beef processors in developing countries may have been reluctant to invest in projects requiring a level of expertise which is not available locally. Another explanation could be that sufficient supplies of beef of the required quality may not have been forthcoming. Other factors may also have contributed to the lack of response to demand for processed beef: attempts at import substitution may have failed as a result of a retailer's bias against locally produced beef products; local producers may not have been sufficiently competitive vis-à-vis imports; or financial institutions may have been reluctant to provide credit for investments in this sector.

The above constraints, which slow down the expansion of beef processing in developing countries, can be overcome if various measures are implemented in order to ensure sufficient supplies of good-quality raw materials and the adoption of appropriate processing techniques by potential meat processors. Some of these techniques are described in this memorandum, including detailed information on a variety of processed beef products which are currently imported by a large number of developing countries.

II. ORGANISATION OF PRODUCTION

A beef processing unit may be operated as an integral part of a slaughterhouse complex or of a butchery business, or as an entirely separate and independent enterprise. Market demand and circumstances determine the type of processing units which may be established in a country.

The integration of a beef processing unit into an existing slaughterhouse complex or butchery business offers a number of advantages. For example, a butcher may supply a segment of the local market with processed beef products made from edible meat trimmings. In this case, the value added may be substantial while additional expenditures may be limited to a few raw materials (e.g. salt, spices, casings) and low depreciation costs for the equipment (e.g. a hand-operated mincer). The extra labour may be provided by the butcher himself or a helper during idle periods. Similarly, the integration of a beef processing unit within a slaughterhouse may be advantageous in view of various economies of scale associated with this arrangement. These advantages do not mean that a beef processing plant operating as an entirely separate business cannot be viable or be as competitive as the other types of plants. A sufficiently large scale of production and demand for specific processed beef products may justify the establishment of a processing plant as an entirely separate business. Such plants operate profitably in a number of developing and developed countries and produce a large variety of beef products.

Although various approaches may be viable, this technical memorandum focuses on the establishment of beef processing units operating as separate businesses. This choice is made in order to facilitate the exposition of the various technical factors and requirements of beef processing and to allow a separate assessment of the viability of a projected plant. However, the information provided should allow, with some adaptation, the assessment of the technical and economic feasibility of processing units attached to a slaughterhouse or butchery business.

III. SCALE OF PRODUCTION, MEAT PROCESSING TECHNIQUES AND RANGE OF PRODUCTS

Meat processing technologies have greatly benefited from recent research and development in food preservation techniques, equipment design and computer-based automation. For example, some of the newly established large-scale plants are equipped with automated machines controlled from a computer room manned by highly skilled operators. Very few workers operate the machines or come into contact with the raw materials between the time these enter the plant and that at which they leave it as packaged meat products. Small-scale producers have also benefited from recent technical developments. For example, it is fairly common for a butchery business in a developed country to make use of meat grinders equipped with a temperature control system in order to avoid the overheating of meat during grinding.

The choice of scale of production is, to a large extent, a function of local and foreign market demand and the amount of beef available for processing. Developing countries which produce large amounts of meat may establish large-scale meat processing plants if most of the production is geared for export. These plants must, of necessity, use capital-intensive technologies in view of the stringent quality control, product uniformity and high level of hygiene required by importers from industrialised countries. In many cases, the need for high technical performance and sophisticated marketing forces some developing countries to allow foreign investments or joint ventures whenever they wish to export processed meat products to industrialised countries.

While the establishment of large-scale, capital-intensive plants may not be avoided whenever production is intended for export, small-scale, relatively labour-intensive units may be preferred if production is intended for the local market. In general, the limited demand for processed meat in developing countries, the high cost or lack of adequate transport facilities (e.g. road infrastructure, refrigerated trucks or wagons) and the limited supply of fresh meat within a given area point in favour of the establishment of such small-scale units. Furthermore, the latter contribute to important development objectives such as employment generation and the improvement of the balance of payments (see Chapter VI). For these reasons, and given the main purpose of the technical memoranda series, this memorandum provides detailed technical and economic information on small-scale beef processing

only. Readers interested in large-scale meat processing plants should obtain
information from equipment manufacturers or engineering firms since the
establishment of these plants require very detailed proprietary information
which is outside the scope of this technical memorandum.

The scale of production described in the following chapters is limited to
a few tonnes of beef products per week. Two plants, with outputs of 1 and 2
tonnes per week, are discussed in detail and are used as illustrative examples
for the estimation of unit production costs (see Chapter V).

The processing techniques described in Chapter III are relatively simple
but do, nevertheless, require electrically powered equipment which is seldom
manufactured in developing countries. In some cases, lower-cost, manually
operated equipment may also be adequate. However, the productivity of some of
the equipment is so low that it may be more profitable to use the more
expensive electrically powered equipment. Potential meat processors need to
assess the alternative pieces of equipment in the way described in Chapter V
in order to identify those which minimise unit production costs. This
assessment takes into consideration equipment and labour productivity, the
prevailing wage level, the unit cost of electric power and the acquisition
cost of the alternative pieces of equipment.

The range of beef products covered by this memorandum includes the
following:

- biltong (an uncooked, cured, dried beef product which is eaten raw);
- charqui (a cured dried product which is cooked before consumption);
- fresh beefburgers (a minced raw beef product which is cooked before
  consumption);
- various types of cooked, cured beef;
- chili con carne (a heavily spiced minced beef product sold cooked and
  almost dry);
- beef frankfurters; and
- beef cervelat.

Production of fresh, chilled or frozen meat is not covered. Similarly,
manufacture of canned meat products is not included since small-scale canning
plants are unlikely to be financially viable.

The techniques described for the seven beef products listed above may be adapted for the production of local specialties which have similar characteristics (i.e. different ingredients may be used or some of the subprocesses may be slightly altered). Thus, information contained in this memorandum could be useful for the production of a fairly large range of beef products.

CHAPTER II

RAW MATERIALS

Meat used in the manufacture of processed beef products includes cuts and trimmings from various locations in the carcase, depending on the nature of the products and the market at which they are aimed. This chapter provides information on the sources and subsequent selection of meat used for the production of cured, salted and heat-preserved beef products. It also describes other ingredients used in these products.

I. SOURCE AND SELECTION OF CARCASE MEAT

Beef is not a uniform commodity. It may be derived from cattle of different breeds, sexes and ages. Its composition and quality may vary according to the type of feed used and the way the animal is handled prior to and at the point of slaughter.

The composition of a carcase, as measured by the proportions of muscle, fat and bone, changes as an animal grows. The relative rates at which these changes take place are influenced by environmental as well as genetic factors. In a young calf, bone might account for as much as 30 per cent of the total carcase weight and commercial yield of saleable meat from such a young animal should be low. Muscle grows relatively faster than bone as the animal matures. Thus, the ratio of muscle to bone slowly increases. Fat makes up only a small fraction of the carcase at birth. It then slowly increases during the first few months of life. Since weight and age are directly related, the weight at slaughter will determine the relative proportions of each constituent of the carcase. Thus, the older and heavier the animal, the lower the proportion of bone and the higher the level of carcase fat and lean meat. Although this is in general true, the sex and breed of an animal will also have a substantial effect on carcase yield.

Some breeds of cattle, in particular the British breeds, begin to fatten at light weights while many of the Bos indicus and other European breeds mature at heavier weights. In general, early-maturing cattle have a lower mature size and enter the fattening phase at lighter weight. Castrated males and young female heifers fatten at lower weights than non-castrated males of the same breed, while old cows may have very high proportions of bone in the carcase due to lack of muscular development and paucity of fat.

In the tropics, cattle are slaughtered at a wide range of body weights and at various stages of tissue development. Although indigenous grading and classification schemes have been developed to differentiate high-yielding beef carcases, such segregation at source occurs in very few developing countries. A meat processor must therefore use experience and a trained eye in order to choose suitable, high-yielding animals and carcases since profitability will depend, to a large extent, on the maximum possible use of the carcase.

Young calves and old cows are generally unsuitable for beef processing unless they can be purchased at suitably low prices. Old bulls tend to have heavy forequarters, but are usually free of excessive fat deposits. They should be a suitable choice for beef processing whenever market forces act to depress their sale price. The most suitable carcase for maximum yield of good quality beef products would be derived from a well-finished mature steer or heifer, lightly covered with surface fat, and having a well muscled carcase, particularly along the loin and in the hindquarters.

II. SLAUGHTER AND CARCASE HANDLING

Careful control of the slaughter operation is essential for the production of good-quality beef products. Although a manufacturer will generally be unable to ensure that he is supplied from well-managed and hygienic abattoirs, careful examination of the carcase will offer clues as to its likely quality.

Carcases smeared with blood, faecal material or dirt should not be used for the manufacture of beef products. Only clean, undamaged carcases free of all offal, apart from the kidney, should be purchased. A manufacturer should also look for marks indicating that the animal has been inspected by a qualified veterinarian.

All meat for onward processing must be certified as free from infectious diseases if any of the products are to be sold uncooked or partially cooked (i.e. when their internal temperature does not reach 60°C).

Cattle that are highly stressed before slaughter as a result of poor handling, or that are exhausted after long road journeys, may produce carcase meat that is dark, dry and firm in appearance. Dark-cutting beef has a sticky texture and water loss during cooking is reduced, thus increasing final product yield. However, this type of meat restricts salt penetration and promotes microbial growth, thus reducing product shelf-life. Carcases having dark-cutting beef should be avoided. However, this condition is difficult to detect in an intact beef carcase. Since the loin is most affected by changes in muscle colour, meat processors are advised to inspect the carcase by making an initial cut between the lumbar and thoracic regions of the loin.

Beef sides or quarters intended for cutting should have been chilled at the abattoir for at least 24 hours in a chill store operating at 2°C. This is necessary if the internal pelvic temperature is to be less than 10°C when the carcase is delivered. In countries where carcases are not ususally refrigerated after slaughter, the meat processor should carefully examine hot beef sides or quarters for signs of taint caused by microbial spoilage. By cutting into the leg muscles overlying the pelvis, it should be possible to detect, by smell, any deterioration caused by growth of bacteria. Chilled sides or quarters that are wet or slimy should also be avoided, since this is indicative of poor handling and ineffective refrigeration.

III. <u>CHOICE OF CUTS FOR BEEF PROCESSING</u>

Trade and economic circumstances influence the supply of meat for the manufacture of processed beef products. In small-scale production by retail butchers, the meat is usually derived from the cheaper parts of the carcase or from trimmings produced by the dressing and cutting of beef. Market requirements generally dictate the nature of the cuts and the quantity of trimmings. Only as a business expands is meat purchased specifically for processing. Many large-scale operators which have no other outlets for the meat process all that they purchase .

III.1 <u>Butchery procedures</u>

Butchery practices for beef cutting are varied. It is not the aim of this publication to recommend any particular butchery procedure. The practice of primal and retail butchery employed will depend, to a large extent, on the way the carcase is split and quartered at the abattoir and the subsequent trade requirements for the retailing of meat cuts. The following brief description of primal butchery is therefore given for the benefit of trained butchers to enable them to identify the primal cuts used in subsequent processing operations. Training in butchery techniques is needed before attempting primal cutting.

The primal and retail cutting procedures outlined in the following account are based on the Pistola cutting system, with the beef side quartered between the fifth and sixth ribs. This particular method has merit since it allows the complete separation of all first quality meat cuts in the hindquarters and loin regions. Figure II.1 illustrates primal cuts and the main muscle components in the beef carcase.

Primal cutting

After separation of the forequarter between the fifth and sixth ribs, the flank piece is removed by freeing the muscles of the abdomen from those of the proximal pelvic limb (see figure II.1). Separation is completed by extending the cut down the side, parallel to and at a distance of some 20 cm from the back bone. The sixth and thirteenth ribs are severed with a saw along the line of division. The flank and rib piece removed by this operation may be trimmed, deboned and used for the manufacture of charqui.

The special hindquarters piece, which contains all of the first quality muscles, may be separated into the hind leg, rump and loin cuts by cutting and sawing directly across the Biceps femoris at a point just below the exposed point of the pelvic bone (see figure II.1). The forequarter is similarly divided into two by removing the foreleg and blade cut from the thorax and neck.

Table II.1 illustrates the distribution of primal cuts within the carcase.

Commercial cutting

Retail or commercial cutting of the final primal carcase cuts should allow a manufacturer freedom to decide which muscles or blocks of muscles could be most suitably used for processing and which, by nature of their retail value, should be sold fresh. Trimmings produced during this operation may be used for the manufacture of beefburgers or chili con carne.

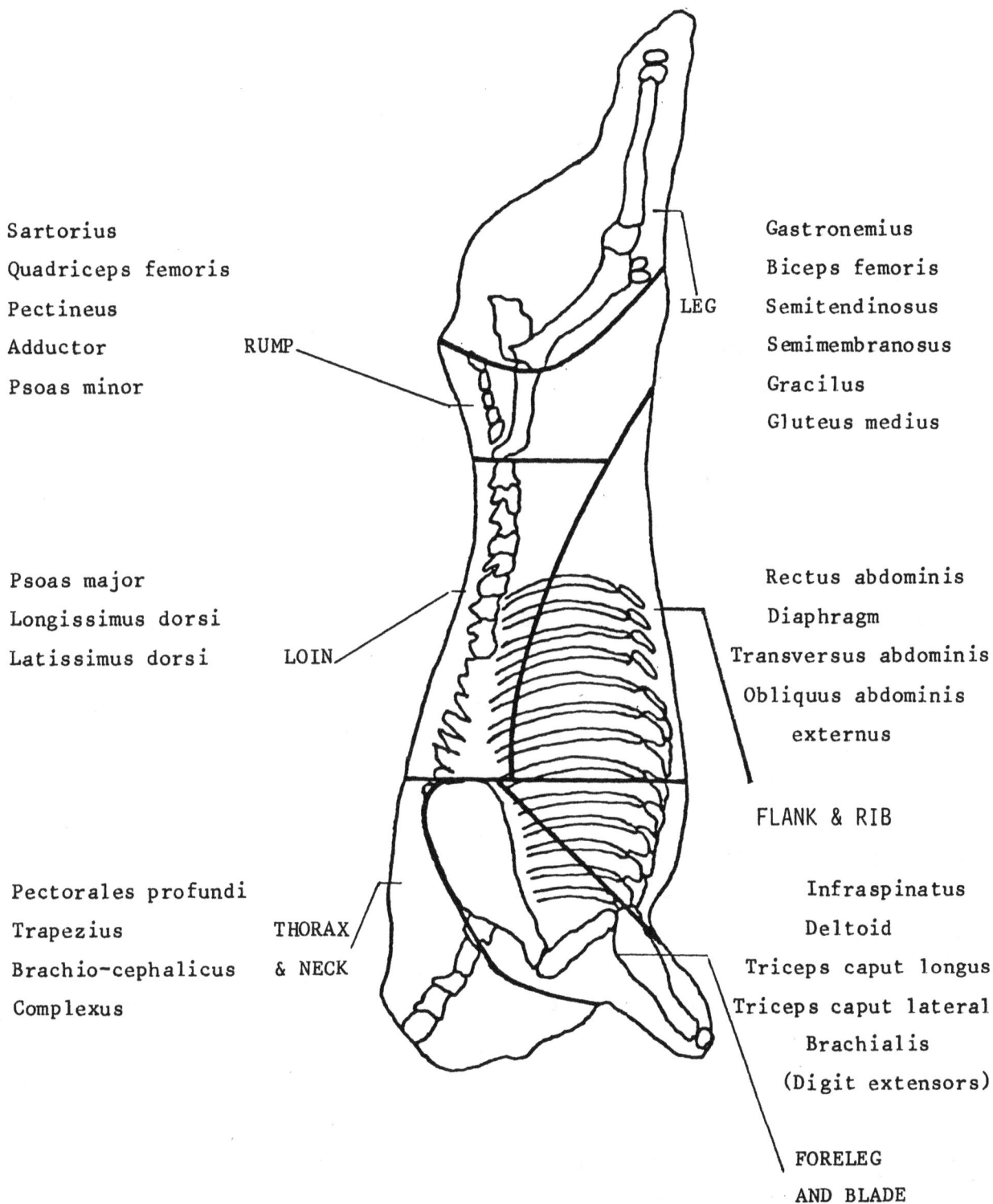

Sartorius
Quadriceps femoris
Pectineus
Adductor
Psoas minor

RUMP

LEG

Gastronemius
Biceps femoris
Semitendinosus
Semimembranosus
Gracilus
Gluteus medius

Psoas major
Longissimus dorsi
Latissimus dorsi

LOIN

Rectus abdominis
Diaphragm
Transversus abdominis
Obliquus abdominis
externus

FLANK & RIB

Pectorales profundi
Trapezius
Brachio-cephalicus
Complexus

THORAX
& NECK

Infraspinatus
Deltoid
Triceps caput longus
Triceps caput lateral
Brachialis
(Digit extensors)

FORELEG
AND BLADE

Figure II.1

Beef carcase showing primal cuts according to Pistola procedure

Table II.1

Primal cuts expressed as a percentage of cold carcase weight

Primal cut	Percentage
Flank and rib	14
Hind leg	29
Rump and loin	20
Foreleg and blade	15
Thorax and neck	22

A butcher might expect the following recovery of lean meat, fat, bone and trimmings from a cold carcase weighing approximately 240 kg:

Commercial cuts (saleable meat)	69.6 per cent
Trimmings	3.6 per cent
Fat	7.9 per cent
Bone	18.3 per cent
Losses	0.6 per cent

Trimmings originating from dissection of primal cuts can be visually assessed for grade and segregated accordingly. "Bought-in" trimmings of unknown origin should be sample dissected into lean and fat.

III.2 Storage and selection of meat ingredients

Carcase meat, cuts or trimmings should be used immediately, or within 24 hours of delivery if held in the chilled state at 0-2°C. Extra care in handling trimmings is necessary. In particular, it is important to maintain their low-temperature state by quick transfer to the chill store. Trimmings and specific commercial cuts (e.g. blade, distal, forelegs and hind legs, neck) which are not for immediate use can be held in a cold store at -10°C or a lower temperature for up to six weeks. Beef trimmings should be compacted and wrapped in airtight polythene bags to prevent the fat from becoming rancid and the meat from losing its colour. Flank or other low-quality cuts of beef intended for the manufacture of salted, dehydrated products are bulky to store. It is therefore advisable to process them immediately without chilling.

The proper selection of meat ingredients is essential for the production of processed beef products of uniform quality. Recognised standards of the gross raw materials with respect to lean/fat ratios and microbiological conditions should be known and consistently met. Additionally, different tissues vary in their moisture to protein ratios, their lean to fat ratios, the relative amounts of pigment they contain, and their water binding properties. All of these properties will affect the characteristics of the end product.

Variations in the price of certain materials will necessitate flexibility in product formulations. While the small producer may need to keep these to a minimum, large producers may substitute raw materials in times of shortage or in order to increase their pofit margins, because they are capable of investigating the effect of these substitutions on the quality of the end-product. Small producers must rely on experience in order to determine the formulations and quality standards which maximise their profit margins.

IV. NON-MEAT INGREDIENTS

A number of non-meat ingredients are used in beef processing. These are briefly described below.

Ice in meat products

Moisture, added as ice at the time of chopping, is important in the formulation and preparation of some beef products. Many of these products would be dry and unpalatable if they contained only the moisture inherent in the meat ingredients. Ice is also necessary to hold the temperature of some comminuted products below that which would make them unstable, and to produce appropriate brine strengths for maximum binding of the meat proteins. Where ice is to be incorporated directly into a product, it is added either in the flake or crushed form.

Salt

Salt is a critical ingredient in the manufacture of many beef products. Apart from its role as a preservative, and in the preparation of meat emulsions, it is important in imparting recognisable product flavour.

Impurities in the salt can cause rancidity in the final product. Thus, heavily contaminated or dirty salt should be avoided. A high moisture content lowers the effective amount of salt present in a given product weight.

Spoilage of the product will occur if this is not taken into account whenever salt is the main preservation ingredient. Storage of salt in sealed containers is therefore recommended.

Nitrates and nitrites

Nitrates (as salpetre-potassium nitrate or chilli salpetre sodium nitrate) and nitrites (as potassium or sodium nitrite) are added to curing ingredients in order to develop the characteristic red or pink colour in the cured meat. In addition, nitrites perform other critical functions. They affect flavour directly or through their action as powerful anti-oxidants that prevent the development of rancidity. Sodium nitrite is also an effective inhibitor of the growth of certain dangerous food poisoning bacteria.[1]

Nitrate is not directly involved in the curing reaction: it is converted to nitrite by a slow process which depends upon the presence of bacteria in the brine and in the meat. Thus, nitrite alone is often used in many rapidly processed products.

The role of nitrates and nitrites in meat curing is currently under review. Nitrites have been implicated in the formation of potentially harmful nitrosamines in bacon and strict limits on their use have been established in many countries. Levels in excess of 200 parts per million (ppm) of sodium nitrate and 100 ppm of sodium nitrite should not be used.

Nitrate and nitrite are added in small amounts. They may be directly dissolved in water if used with brines. For dry cures, they may be initially dispersed in the salt component to ensure uniform distribution. Nitrite salt, used for rapidly produced products, consists of 99.5 per cent salt and 0.5 per cent sodium nitrite. It is made up on a daily basis.

Sugars

Sugars, such as sucrose and dextrose, are used in the manufacture of some sausages to improve product flavour and colour, mask saltiness and, in some cases, to bring about necessary chemical and microbial changes in the product.

---

[1] Sodium nitrite is added to inhibit the growth of Clostridium botulinum, a food poisoning organism which produces a fatal toxin.

The levels of sugar added are limited by regulation in some countries. Amounts in excess of 2.5 per cent by product weight are not recommended. Artificial sweeteners, such as saccharine, are used in some countries at a level of 0.01 per cent to improve the flavour of some cured products.

Anti-oxidants and mould inhibitors

A number of products may be added to fresh beef products to retard the development of oxidative rancidity. The most commonly used products are butylated hydroxyanisole (BHA) and butylated hydroxytoluene (BHT). They may be used at levels between 0.002 and 0.003 per cent of the finished product weight, depending upon fat level in the finished product. Other food additives may be used as thickener, anti-oxidant, colour or flavour enhancer and so on. A list of additives, including permitted utilisation levels, is provided in Appendix I.

Spices and seasonings

Spices, aromatic substances of vegetable origin, herbs and vegetables are used to season beef products.

Spices may be added as a natural spice or as spice extract. The extracts are easier to store because they are less bulky than the natural product. Most spices contain moulds and bacteria which can cause considerable damage to products under conditions favourable to their growth. Some spice manufacturers sell "purified spices" where the level of contamination is greatly reduced. However, the purification processes are not totally efficient and some risk remains in their use.

Spices, herbs and vegetables should be stored in a dry goods store, preferably in sealed containers. They should not be exposed to direct sunlight or steam and should only be ground or crushed on the day of manufacture.

Sausage casings

There are two main types of casings used in the manufacture of processed beef products: animal casings (also called natural casings) made from the stomach, intestines and bladders of cattle and sheep, and synthetic or manufactured casings which may be made from cellulose or collagen and which are either edible or non-edible. The processing of animal casings is labour-intensive and involves a considerable amount of cleaning and stripping away of unwanted layers. It should not be carried out by untrained personnel.

(i) <u>Natural casings</u> : In the production of natural casings, high yield and good quality depend upon the complete removal of fat, nerves, blood and lymphatic vessels without damaging the remaining tissues. The intestines are removed without puncturing and their gross fat is trimmed off. They are then detached from the mesenteric tissue into strands (termed "running"), and cut into appropriate lengths and types. The contents of the intestines are removed by stripping either by machine or by hand, under warm water, by easing through a restricted aperture in the case of small casings and by turning them inside out and massaging with large casings. Small casings are then "slimed" by repeated passing through rollers and strippers or by hand working after a prolonged soaking in lukewarm water (one to six hours depending upon source) in order to remove the mucosa and muscle layers. Large casings are passed through crushing rollers inside out and brushed (rice root and bristle) under a hot spray. Hand treatment is by careful brushing under warm water.

Washed casings are stored overnight in cold, 20 - per - cent saturated salt solution. Their quality is checked by inflation and visual inspection for scars and perforations. They are subsequently rubbed with medium dry salt and left for one week to cure. Excess coarse salt is then removed and fine salt is rubbed in. They are finally packed (usually in a 55 gallon barrel or a tierce) and held at 5°C until use. It is usual for the user to recheck the quality on immediate arrival from the supplier. The variety and uses of animal casings are shown in table II.2.

Natural casings are generally received by the beef processor in lengths of 80 to 100 metres, either dried or salted. Ideally, they should be stored at 5°C. At any rate, they must never be frozen or exposed to heat. If allowed to absorb water, they will spoil and become worthless. Casings stored dry at 5°C can keep for several weeks. If possible, casings should be used the same day they are removed from salt. Any left-over casings must be re-salted even if they are for use the following day.

(ii) <u>Manufactured casings</u>: The use of manufactured casings has increased significantly in recent years as a result of demand for uniformity of stuffed diameter, ease of handling and high resilience to breaking.

Table II.2
Animal casings and their use

Variety	Description	Uses
Sheep casings or "strings"	Small and large intestine, graded according to un-stuffed diameters	Used for top quality frankfurters, cocktail sausages, fresh sausages
Beef rounds	Small intestine	Metwurst, Bologna sausage
Beef middles	Large intestine	Many 'continental-type' sausages
Beef weasands	Oesophagus from throat to first stomach	Bologna sausage, luncheon meats
Beef bungs	Appendix vermiformis	Large cooked sausages
Beef bladders	Bladder	Mortadella and French garlic sausage

Cellulose casings, produced from cotton, may be grouped into three types:

(a) Small cellulose casings are used for the manufacture of frankfurters. They may be purchased either clear or coloured, in lengths ranging from 25 to 55 metres.

(b) Large cellulose casings are available in three types: regular, high-stretch and light-weight casings. Large regular casings are used for many types of processed beef products and act primarily as a container. High stretch casings are specially treated to give the casing extra stretch and shrink characteristics.

(c) <u>Fibrous casings</u> have high-strength characteristics and are used in cases where a maximum uniformity of finished product diameter is desired. They are therefore ideal for products which must be sliced into pre-packaged commodities. Several types of fibrous casings are available, differing in their treatment and the uses to which they are put. Moisture-impermeable fibrous casings are used to improve the keeping quality of certain products, but they are also impermeable to smoke: they should only be used for water-cooked items such as liver sausage. Special fibrous casings which adhere to the meat contents during dehydration and shrinkage are also available. These are particularly useful in the manufacture of dried products such as cervelat.

CHAPTER III

SMALL-SCALE PLANT AND FACILITIES

The type and extent of plant and machinery necessary for the manufacture of beef products depend upon the size of the proposed operation and the variety of products to be processed. Before entering the processed meat trade, the processor must decide what range and quantity of products he is going to manufacture, bearing in mind his estimates of demand and of the available resources (capital, raw materials, etc.).

The processor must seek to add high value to his processed commodity. On very small throughputs, this often means specialising in a single product from a particular raw material component, in addition to the sale of fresh meat. Therefore, very small-scale production is often associated with a small retail butchery business.

For illustrative purposes, this memorandum considers the principles of plant operation, buildings and equipment needed for two small-scale units producing a small selection of products (see Chapter IV). Both units utilise forequarter and flank areas of the carcase only, with specific areas of the carcase being utilised for the particular end products for which they are best or only suited (e.g. flank and rib for charqui). These two units (designated as Model 1 and Model 2) are only examples of possible product mixes. They do not represent recommendations as to the most appropriate production quantity or mix which must be adopted by a small-scale unit. These can only be determined in the light of specific local market situations. These model operations are used in Chapter V to calculate fixed capital and annual operating costs.

The weekly production profile for Model 1 is as shown in table III.1:

Table III.1

Weekly production profile for Model 1

Product	Raw material source	Processing steps	Amount (kg) per week
Beefburgers	Neck, foreleg, trimmings kidney knob, channel fat	Minced, shaped, fresh	500
Chili con carne	Blade, shoulder, neck (lean), kidney knob, channel fat	Minced, cooked, filled	375
Charqui	Flank and rib	Dry salt, dry (repeated)	125
Total			1 000

Model 2 is basically the same plant as Model 1. It has the same product profile but the weekly output is increased to 2,000 kg per week by working two shifts instead of one.

I. PLANT LAYOUT

Although a plant layout depends to a large extent, upon building shape and floor area, the simplified layout suggested in figure III.1 embodies principles of construction and design that should be followed as far as possible whatever the other limitations might be.

The movement of product should try to follow, whenever feasible, an unbroken path from the receipt of the raw materials through the various preparation and processing operations to the area of dispatch.

In large units, conditions of hygiene require the physical separation of areas set aside for fresh, cooked and cured meat product processing. This is not possible in very small plants of limited space. In this case, every effort should be made to ensure that fresh meat and cured meat are not being processed concurrently, and that fresh and cooked meats do not come into contact.

In the scheme set out in figure III.1, beef forequarters or primal cuts are unloaded, weighed and inspected before being designated to commercial cutting, immediate processing or cold storage. Flank pieces, when delivered separately, are transferred directly from the delivery platform to the salting area. Primal and commercial-cut butchery operations are carried out in area A. Trimmings or specific low-quality primal cuts should be immediately removed for the manufacture of beefburgers. All other deboned cuts should be returned to the 0°C cold store or deep-freeze cabinet to await processing. All remaining beef processing operations should be carried out in the main processing hall (area B) or in the drying room (C). Although the processing operation should ideally be carried out under refrigeration, the refrigeration plant necessary to cool the entire processing unit would represent a large capital investment which cannot be justified for small-scale operations.

All cooking and smoking operations should be carried out in a separate room to facilitate control of steam and waste-water disposal and to prevent the temperature from rising in the processing area.

I.1 Raw material storage and refrigerated curing rooms

The design and operation of cold storage facilities require professional expertise, especially in tropical environments where problems associated with the extraction of heat in high ambient temperatures are compounded by condensation from high humidity levels. The advice of a local expert must be sought at the project evaluation stage.

Raw materials, in the form of beef sides or quarters, should be stored in a walk-in cold store operating between 0°C and 4°C. The store should be located as close as possible to the reception area, and all deliveries should be carefully examined, weighed and date-coded. Calculation of the size of store depends upon the delivery of raw materials and production sequence. For small-scale units, deliveries are usually made on a daily basis.

Carcases must hang in a way which allows sufficient space between them for personnel to move around. The calculation of hanging room dimensions for the two models under consideration takes account of the carcase and rail-spacing criteria indicated below.

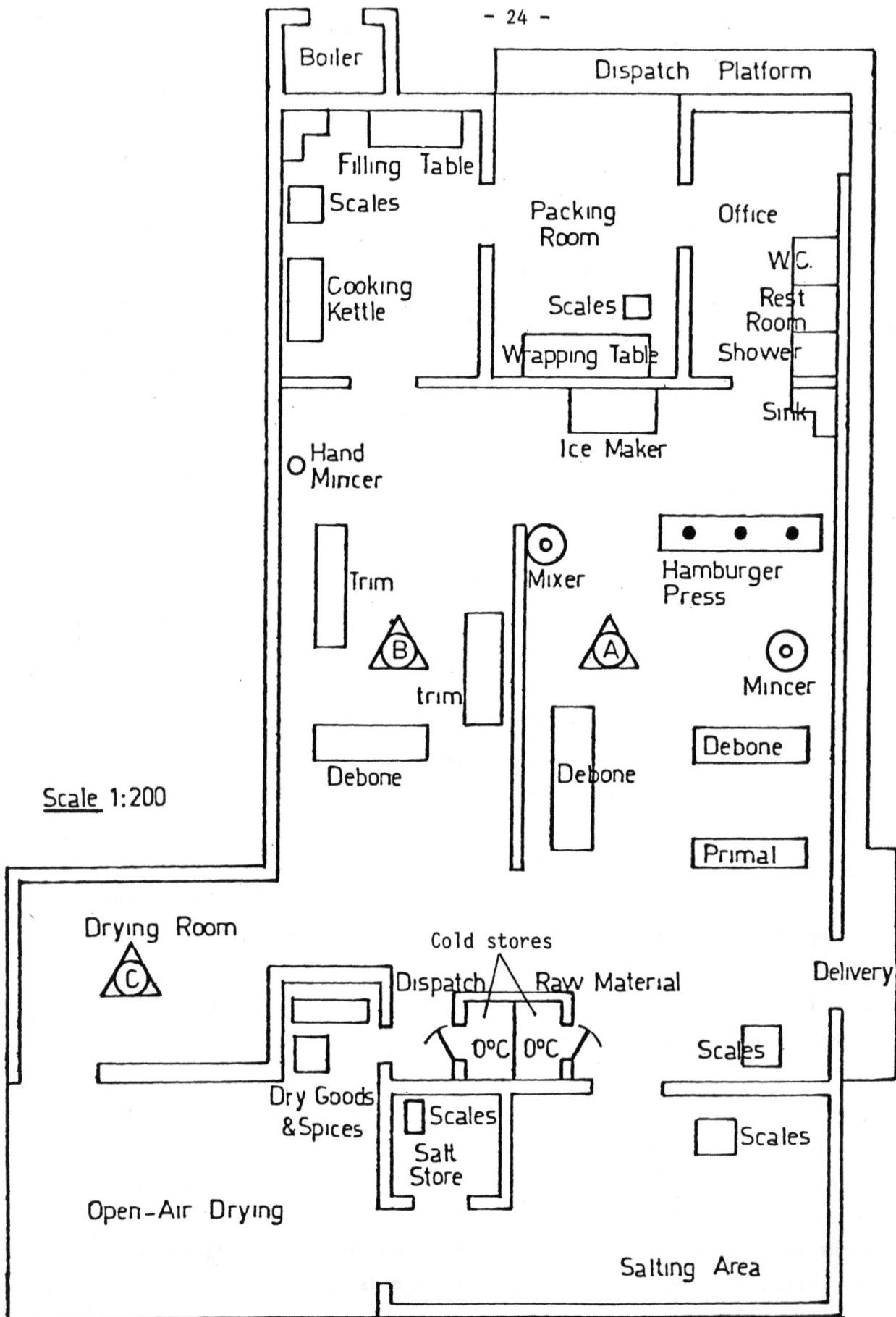

Figure III.1

A schematic floor plan for small-scale beef processing plant

(Model 1)

| | Distance between rails (metres) | | Minimum distance between |
	Minimum	Desirable	carcases along rails (metres)
Beef	0.9	0.9	0.20
Lamb	0.46	0.46	0.15

There should be a gap of at least 20 cm between carcase surfaces and a walking space between cold-store walls and carcase meat to facilitate movement. For very small cold stores, the area per forequarter is estimated at 0.57 m^2. Taking into account a movement factor of 1.4, the actual area per forequarter (FQ) is equal to 0.80 m^2 (i.e. 1.4 x 0.57).

The above factors may be used to calculate the raw material cold-store area for Model 1 as follows:

- carcase meat demand (bone-in): 273 kg/day (estimate based on recipes for beefburgers, chili con carne and charqui provided in Chapter IV);
- flank diverted to salting operation: 104 kg/day;
- carcase meat (FQ only) for storage: 169 kg/day;
- assumed average weight per FQ: 45 kg;
- number of FQs required: 4;
- cold store area requirement: 4 x 0.8 m^2 = 3.2 m^2.

This area is shown in figure III.1 as a cold store area of 2.0 m x 1.6 m.

The raw material cold store should have an operating temperature of 2°C with a tolerance of $\pm$ 2°C. The relative humidity should be maintained between 85 and 90 per cent.

Materials, such as trimmings and fat, that are not to be used within 24 hours of delivery should be housed in a chest deep-freeze until required. Defrosting should be carried out by placing the frozen materials into the raw materials (0°C) store 36-48 hours prior to their processing. It is important to date-code all meat that is to be frozen.

A refrigerated curing room will be necessary in the manufacture of brine-cured products (i.e. brisket). Space must be allowed for the raw materials, brine and container. Since curing processes take more than one day, allowance must be made for the number of days in cure. The preferred method of processing brined products in very small units makes use of small containers (30 litres) and employs a packing system within the cold store.

Total product occupancy volume can be calculated by assuming a "stackable" height of 1.8 m. The product occupancy area can be calculated by dividing the total product volume by the assumed height. This area is then multiplied by a movement allowance factor of 1.2 in order to obtain the cold-store area. The height of the cold store may be calculated by multiplying the assumed stackable height by 1.25 (allowance factor for rail supports, etc.).

Specialist advice should be sought regarding refrigerative capacity, the determination of which depends on a variety of factors within the cold store as well as climatic conditions. Information on size, product flow and cold-store distribution need to be collected for the estimation of cold-store volumes by a refrigeration engineer.

I.2 Finished products stores

All finished, cured beef products should be held under refrigeration prior to dispatch. They should be kept separate from all unprocessed raw materials. Care should be taken to ensure that they are properly packaged and date-coded. The size of the store depends upon the system of marketing employed. It should, however, be sufficient to obviate dispatch difficulties.

Uncured, fresh beef products such as beefburgers should also be stored in a cold store at $0°C$ prior to despatch if they are not being marketed immediately after processing.

I.3 Dry goods store

A room for the storage of spices and processing ingredients not requiring refrigeration, for processing equipment and for packaging materials should be included in the plant layout. The size of the room is directly related to the number of products processed and, more importantly, to the frequency of delivery of spices and other ingredients. It must be remembered that spices and colourants tend to lose their properties with age, and that the storage period should, therefore, be as short as possible.

This room may also be used as a weighing area for curing ingredients and spices.

I.4 Cooking and smoking room

The cooking room, which should be of generous dimensions and separated from the main processing area by self-closing doors, should contain all the equipment used in the cooking and smoking of the products. It must be well ventilated to ensure the exit of steam and smoke, have adequate drainage and be served with treated cooling water. The steam boiler (and generator/compressor) may be located in a separate building close to but separate from this area.

I.5 Main processing hall

The layout of the main processing hall should facilitate uninterrupted product movement and eliminate the need for excessive product handling. Table-mounted mincers, mixers and stuffers should be sufficient to cope with the limited throughputs of small-scale units. The tables should be of stainless steel. Carcase butchering should be carried out on high-density polythene cutting blocks instead of traditional wooden blocks which are less hygienic and more difficult to keep clean.

The size of the processing hall may be estimated on the basis of the area occupied by the various pieces of equipment and of an assumed access circumference of 0.5 m multiplied by a movement allowance of 1.4. This estimation procedure applies, in particular, to small-scale meat processing plants. Large units (e.g. 75 tonnes per day) utilising large pieces of equipment tend to have a greater product movement area (e.g. a movement allowance factor of 1.6 to 2) in order to promote the maximum efficiency of labour.

The processing room should be constructed in materials that may be easily cleaned and yet provide a safe and pleasant working environment. Particular care must be given to lighting, drainage and service facilities and to the provision of a non-slip floor.

II. PROCESSING EQUIPMENT

A minimum list of meat processing equipment should include simple equipment for the mincing and mixing of meat components and that needed to smoke and cook the finished product. Equal importance must be given to the range of tables, trucks, scales, bins and other accessories necessary for a fluent operation. Having decided the range and quantities of products required, the processor must choose among alternative processing techniques. The cost and skills of local labour may influence the system chosen. In particular, the processor may have the option of choosing either a labour-intensive system utilising less sophisticated equipment or a capital-intensive system involving larger initial investments but lower labour costs. The equipment suggested for the two models described in this memorandum should, however, be suitable for a wide range of conditions.

Appendix II provides estimates of the productivity of labour and of various pieces of equipment for the main meat processing operations (e.g. mincing, mixing, stuffing). These estimates may be used to determine the size and number of pieces of equipment required for a given scale of production, taking into consideration the number of shifts per day and the number of working hours per shift, as well as the sequence and duration of processing stages. Some pieces of equipment (e.g. mincers) will be used full time while others will be used only part of the time. Consequently, the capacity or rate of the former pieces of equipment will determine the scale of production.

In units producing a variety of products, flexibility is necessary to meet demand changes in particular products. Equipment installed for operation by a single worker should be operable by more labour in times of increased demand. For instance, installed equipment for a 5-kg batch of meat products may have an operating daily throughput of 58 kg for single-worker operation and 225 kg for 3.9 workers (although this does not include additional tables, and so on). If beefburgers are only one of a variety of products, equipment output can be varied considerably by mobility of labour within the unit. Thus, for Model 1 (for which installed equipment corresponds to a 20 kg-batch capacity), the entire unit output could be transferred to beefburgers production by reallocation of labour.

Equipment for use in the tropics must have the following characteristics: to be made of appropriate materials to avoid corrosion; rugged construction to minimise maintenance; satisfactory design to limit the handling of meat and to facilitate thorough cleaning after use; and the capacity to meet any expected production requirements. All equipment should be provided with the necessary safeguards to ensure that it is not hazardous to operators. The most important pieces of equipment of interest to small-scale meat processing units are briefly described below.

II.1 Mincers

Mincers are used to cut meat into small pieces so that they may be thoroughly mixed with other ingredients or curing salts. During mincing, meat is fed from the hopper to the mincing plate by means of an auger. As the meat is extruded through the holes in the plate, it is sliced by revolving knives. Plates with holes of various sizes are available, depending upon whether the meat is to be cut into large or small pieces. Unless mincer plates and knives are kept in good condition, meat will be heated during mincing and will lose quality. Power costs will also be increased and delays experienced in passing the material through the mincer.

Hand mincers are available in capacities up to 50 kg/hour. However, all hand mincers require that the raw material be previously reduced to portions of approximately 100 mm by 50 mm. In addition to their higher throughput, powered mincers can process larger pieces of raw material, and thus save valuable butchers' time.

Powered mincers are available in varying sizes, from bench or table-mounted, single-speed models with capacities of 80 kg per hour, to automatic, double-feeding, dual-speed models capable of handling pre-broken frozen meat up to 6,000 kg per hour (these mincers are often referred to as grinders). A manufacturer should choose a machine whose quoted output is higher than production requirements since these continuously rated outputs are often only realised under ideal conditions. A mincer suitable for small-scale beef processing operations is shown in figure III.2.

II.2 Mixers/blenders

Mixers and blenders, although often used for the same task, are distinctly different in the way they function. Mixers simply mix the product to incorporate all of the ingredients. Blenders, on the other hand, mix the product and perform the necessary mechanical agitation required for product binding.

Figure III.2

Mincer for small-scale beef processing

Mixers are round-bottomed tanks equipped with wing-shaped paddles revolving in opposite directions. They vary in size from 10-litre bowl capacity, bench-mounted models (see figure III.3) to 7,000-litre models fitted with pneumatically operated discharge hatches and bucket-lifting hoists. The small mixers with a bowl capacity of 10-30 litres are suitable for small-scale processing operations.

Some companies supply equipment where the mincing and mixing functions are incorporated into one single machine. This equipment facilitates production and eliminates the need for manual manipulation of the minced ingredients.

II.3 Cutters/bowl choppers

In a cutter or bowl chopper, comminution and mixing of the meat ingredients are accomplished by revolving the meat in a bowl past a series of knives mounted on a fixed shaft that rotates at high speed. The meat is guided to the knives via a plough-shaped arrangement fitted on to the side of the bowl (see figure III.4). In sophisticated cutters, bowl rotation, knife speed and temperature may be controlled and vacuum hoods fitted to enable chopping under vacuum.

To ensure correct cutting, the knives must be carefully adjusted and set as instructed by the manufacturer. The action of the knives raises the temperature of the meat. If the latter is allowed to become too warm, the characteristics of the emulsion will be lost. Cutting time should be regulated through careful control of temperature or, in the case of beefburgers manufacture, by examining the size of meat particles.

Bowl choppers range in size and complexity from a 10-litre, single knife model to a 500 litre, multiple-knives vacuum model. There are various combinations in all capacities. The purchase of this piece of equipment should be made with great care since the ease of product manufacture depends upon it. Single-knife choppers are satisfactory for the single product manufacturer, especially where the cutting process can be interrupted to allow for additional cooling. For the production of a variety of chopped products, a multiple blade, two-speed instrument is essential.

In all cases, the machine should be demonstrated under production conditions before a final selection is made.

Figure III.3

Mixer suitable for small-scale beef processing

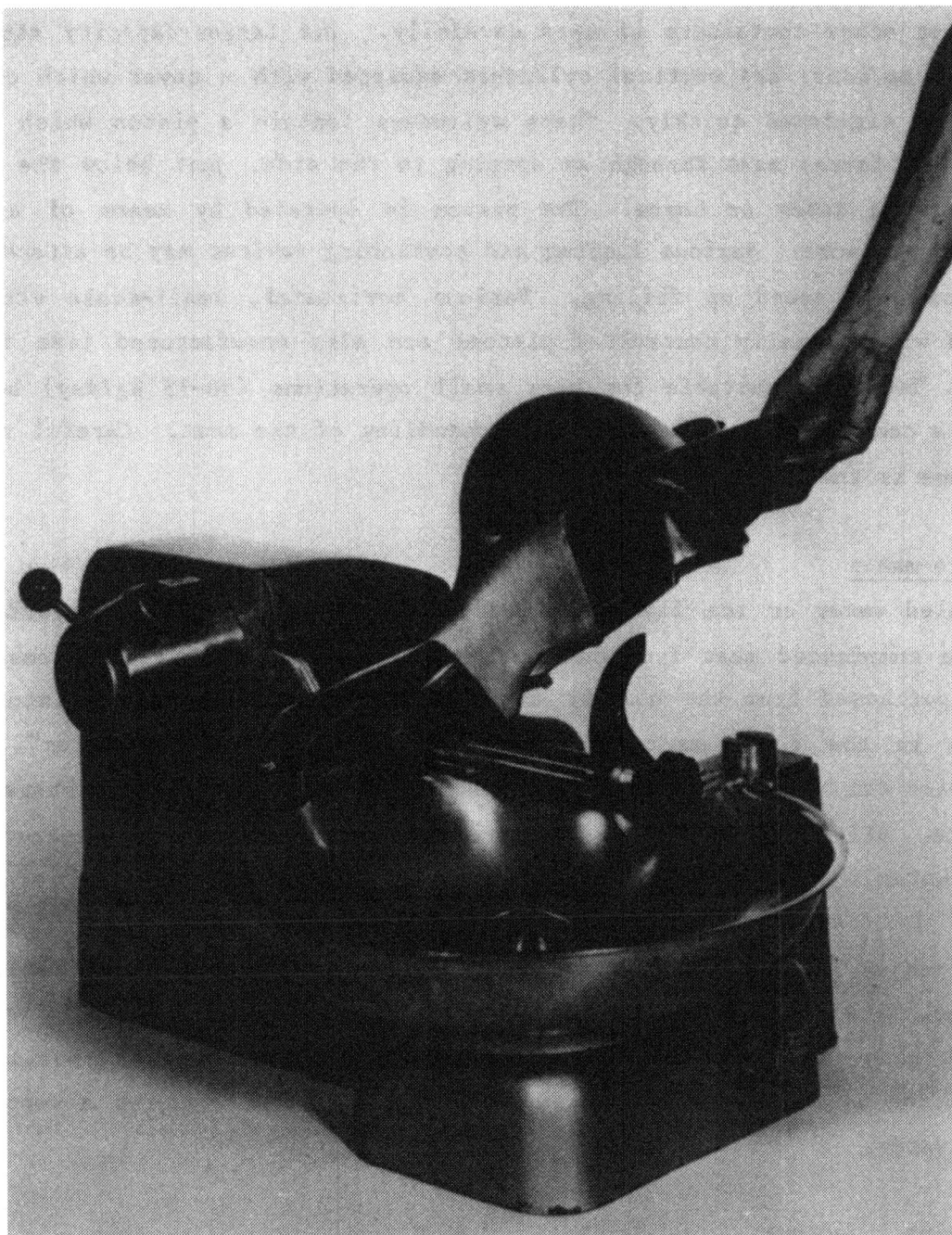

Figure III.4

10-litre capacity bowl cutter, with single blade

II.4 Extruders/stuffers

There are two types of extruders: the pump type and the stuffer type. The pump-type extruder is usually used for high-volume (1-5 tonnes/hour), fine-cut products. It cannot, therefore, handle coarsely chopped emulsions such as beefburgers. Piston stuffers fill all types of meat mixtures into casings or other containers if used carefully. The larger-capacity stuffers (e.g. 250 kg/hour) are vertical cylinders equipped with a cover which can be removed or tightened quickly. These cylinders contain a piston which moves upwards and forces meat through an opening in the side, just below the cover into stuffing tubes or horns. The piston is operated by means of air or hydraulic pressure. Various linking and portioning devices may be attached to the opening to speed up filling. Various horizontal, small-scale stuffers equipped with manually controlled pistons are also manufactured (see figure III.5). These are suitable for very small operations (50-75 kg/day) but do involve a considerable degree of manual handling of the meat. Careful regard to hygiene is therefore essential.

II.5 Ice-maker

Chilled water or ice flakes are added at the cutting stage in order to keep the comminuted meat from overheating. For small-scale operations, ice can be purchased from the nearest ice plant. It is incorporated into meat products in the flake or crushed form to maximise "cold transfer". Any surplus ice can be used for cold brine make-up, thus saving on cold store duty and space. All ice used in meat processing operations should be prepared from treated water.

Ice-making plants with capacities of 70 kg per 24 hours and higher are available. All of them incorporate stainless steel lined bins to hold the ice and most plants are fitted with automatic controls to start ice-making and shut off the unit when the bin is filled. The ice-maker requires a supply of treated water.

II.6 Brine injectors

Single-needle or multi-needle brine injectors are used by the meat processing industry to pump brines of any desired concentration into meat tissues. Extremely good brine distribution may be achieved by relatively unskilled operators. The simplest type of injector, such as that shown in figure III.6, uses a single needle with a number of holes along its length. The machine is designed for hand operation. Automatic, multi-needle injectors

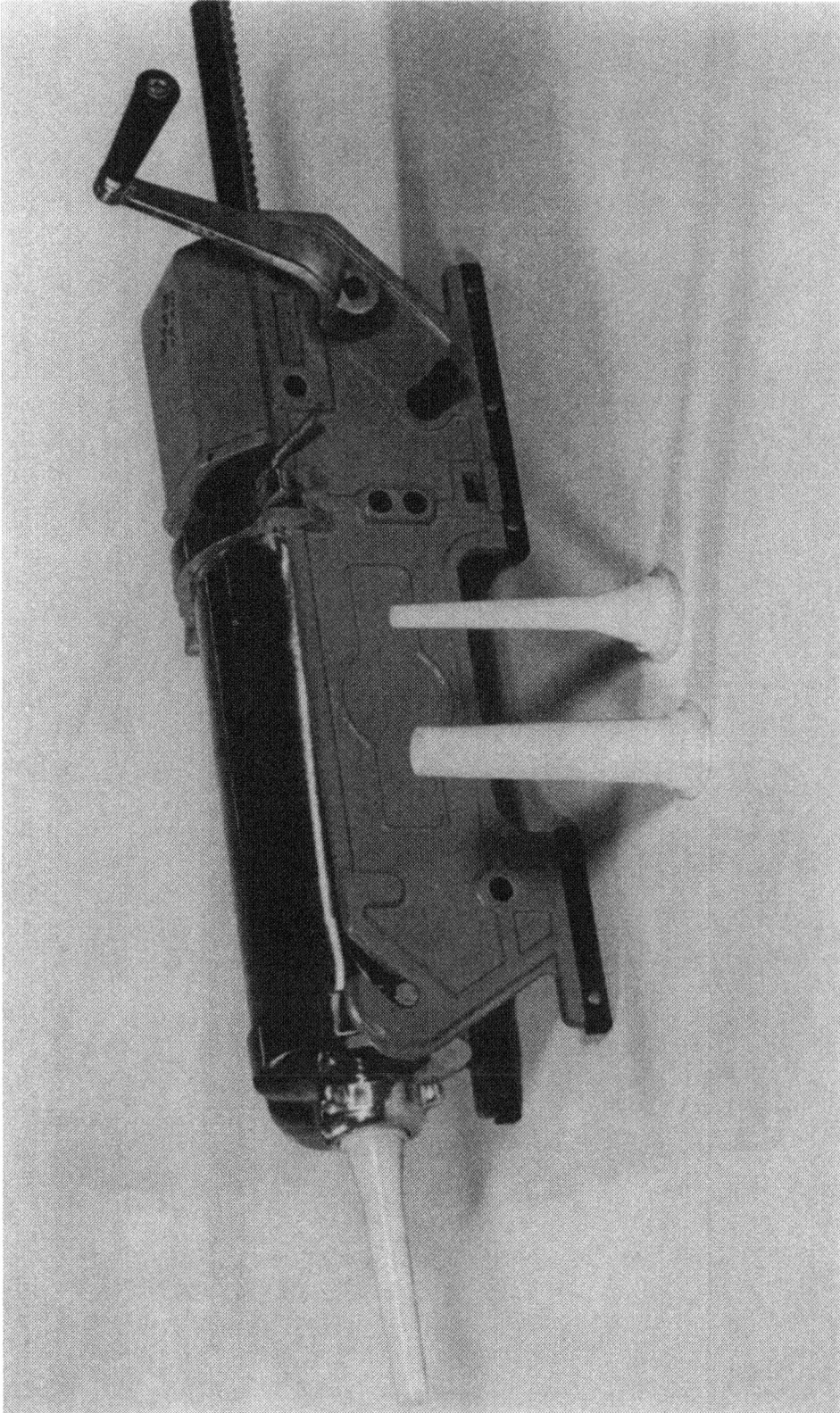

Figure III.5

Horizontal stuffer, manually operated

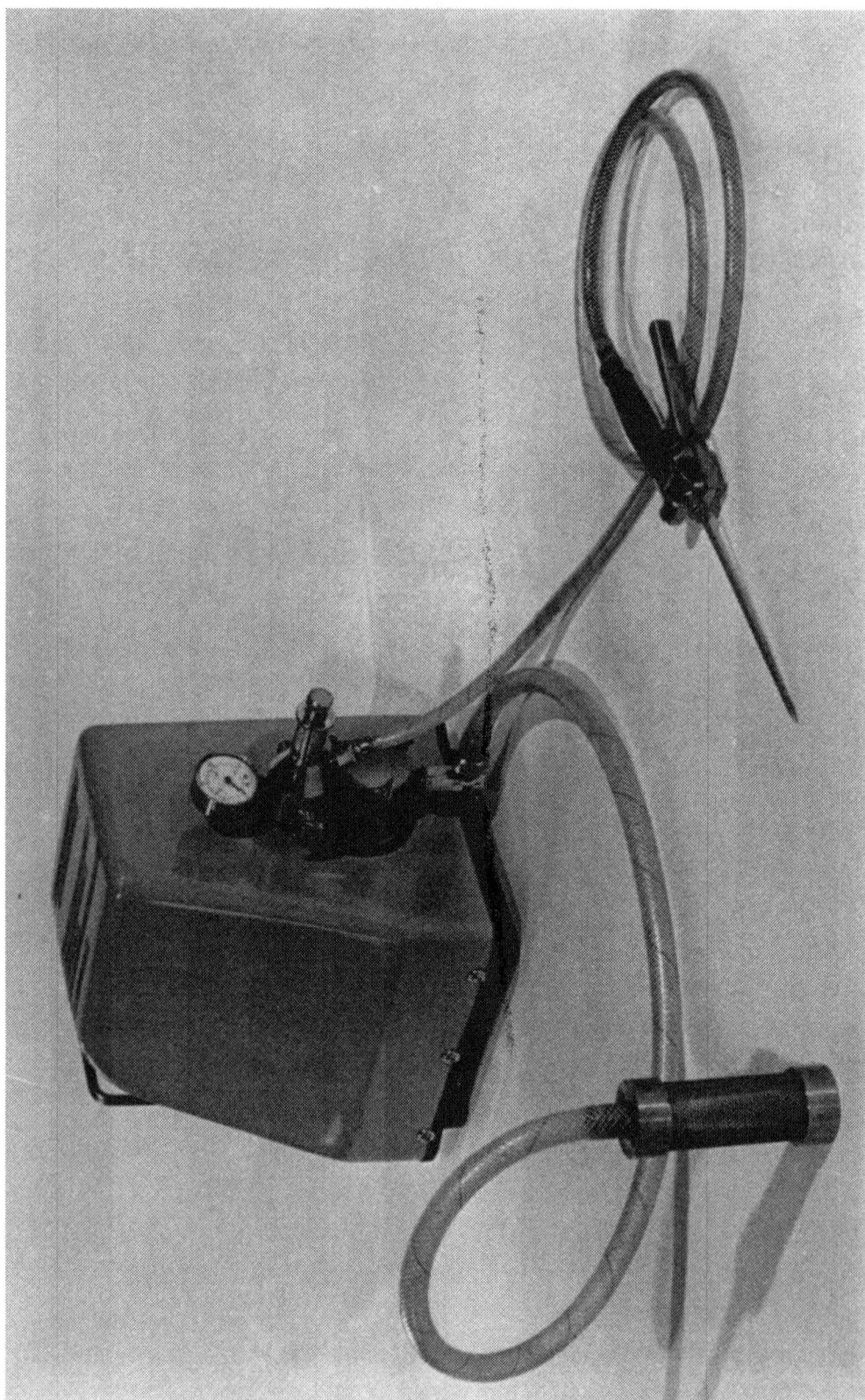

Figure III.6

Single-lance brine injector

offering the processor an opportunity for rapid processing are also available. These models would not be suitable for small-scale operations. Brine injection systems can be used in combination with dry salt or wet-soaking curing procedures.

II.7 Smokehouses/processing ovens

In addition to providing a chamber for the smoking of meat and meat products to an acceptable colour and flavour, modern smokehouses/processing ovens are also designed to perform the critical process of heating and cooking.

Control of temperature rise in products is very important. For example, lean tissue must receive sufficient heat to develop the red cured meat colour, yet the temperature must not be allowed to rise high enough to oversoften the fat. Failure may result from a deviation of 3°C to 5°C from the theoretical processing temperature. The smokehouse temperature must be limited to a few degrees centigrade about the final internal meat temperature.

Smoking and cooking operations are under continual review and a number of improved plant designs are now commercially available. Most modern plants employ smoke generators, comprising a box or drum equipped with a mechanical agitator and blower to draw smoke across the product. The simplest possible system consists of a pile of damp hardwood sawdust which is burnt under or alongside the smokehouse. Another type of smoke generator is also available. It uses dry sawdust which is fed into a small, electrically heated chamber and burned in air to an ash. This type of generator yields a high volume of smoke for the amount of sawdust used but requires more maintenance than the wet sawdust model.

Figure III.7 illustrates a simple type of smokehouse with a working volume of 1 m³ (equivalent to 50 kg of sausage product) which could be built from local materials. Its performance as a cold smoking unit could be improved by the addition of an internal fan, thereby creating convection currents within the unit. Thermostatically controlled electric heating can be installed to enable the kiln to be used for hot smoking (i.e. partial cooking). Cooking can be completed in cooking kettles. Products are moved to and from the unit by mobile trolleys which may also be produced locally.

Specialist products, such as whole beef frankfurters, require a more sophisticated control of smoking and cooking procedures for consistent quality. In this case, the cooking and smoking operations are combined in the

Smoke
Exhaust
Duct

Flue
Control

Fire
Boxes

Smoke
Stick

Kiln
Trolley

Smoke
Diffuser

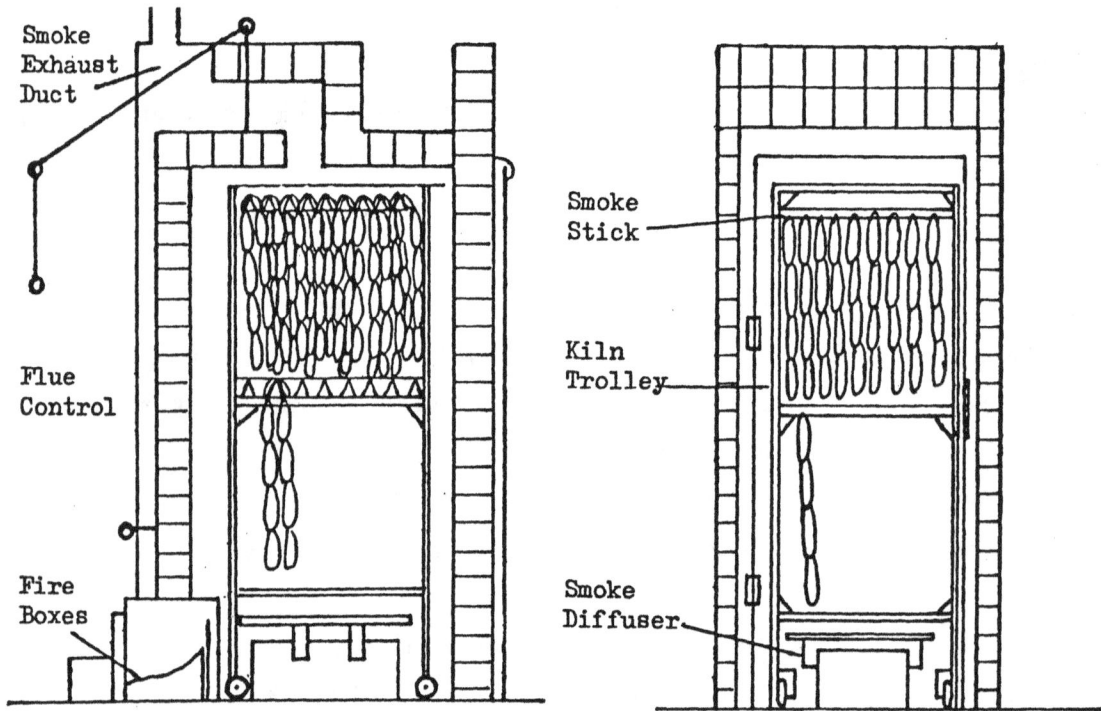

Figure III.7
Simple masonry type smokehouse

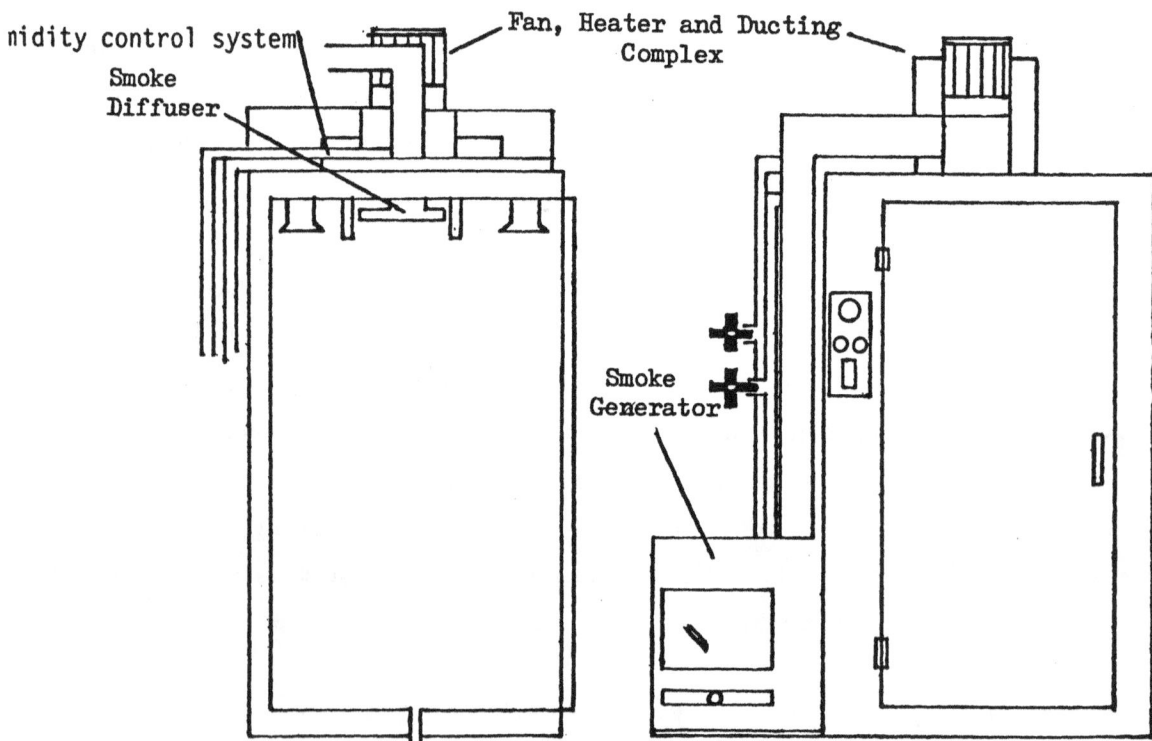

nidity control system

Smoke
Diffuser

Fan, Heater and Ducting
Complex

Smoke
Generator

Figure III.8
Air-conditioned smokehouse and processing oven

same equipment. This type of air-conditioned smokehouse is illustrated in figure III.8. The regulation of temperature, humidity and volume of circulated air and the density of smoke is accomplished by varying the amount of outside air or by injecting steam into the house. This type of smokehouse (or processing oven as it is often called), can perform, in addition to smoking, all the conditioning and cooking functions, including cold or hot showering.

It should be noted that smokehouses using natural smoke generators require appropriate air pollution control equipment.

II.10 Other cooking equipment

There are many types of cooking equipment used in beef product manufacture apart from the multi-purpose processing oven. Products such as chili con carne are often cooked in steam-jacketed, round-bottomed kettles made of stainless steel in order to prevent burning (see figure III.9). Live steam or electrically heated kettles are also available.

Whole muscle products such as silverside or brisket are often cooked in cookers used for moulded products. Product load and temperature are closely controlled to minimise losses. Many sizes of moulds and cookers are available.

III. PACKAGING MATERIALS AND EQUIPMENT

Good packaging is an essential element in the manufacture of high quality beef products. A package must not only protect goods during transport and handling but also help to preserve good appearance and freshness throughout retailing.

The packaging requirements of fresh, cured and processed beef products differ considerably. These requirements are briefly reviewed below for fresh products and cured meats respectively.

III.1 Fresh beef products (e.g. beefburgers)

Beefburgers rely on oxygen for the development of a bright red colour. They are usually sold wrapped in oxygen-permeable films such as cellophane, polyvinyl chloride and polyethylene. Prolonged storage at refrigerated temperatures in these films will however result in irreversible colour and flavour changes as well as microbial spoilage. For long-term storage, wrapped beefburgers should be packed into cardboard boxes and frozen at -18°C.

Figure III.9

Steam-jacketed, round-bottomed kettle

III.2 Cured beef products (e.g. cured silverside or brisket)

The pigments in cured beef products are stable only in the absence of oxygen. It is essential, therefore, that these items be packaged in low oxygen transmitting material. Sliced products are particularly affected by exposure to air and light. Packaging in low oxygen permeable films such as polyester, polyamide and vinylidene chloride is acceptable for short-term storage of cured products where vacuum packaging machinery is not available. Shrink films and stretch films improve product appearance and are generally preferred to simple bagging or over-wrapping methods. For long-term packaging of cured beef, vacuum packaging is recommended. Vacuum sealing of products in bags or pouches made of plastic laminates (e.g. nylon and polyethylene) may be carried out in vacuum chambers or by the more simple evacuation and clip sealing machine. The latter consists of a screened vacuum pump working through a clipping nozzle. The material to be packed is enclosed in a laminate outer, presented to the nozzle and the vacuum established. This is accomplished within five to 30 seconds, depending on the air content of the pouch. On completion of evacuation, the pouch is clip-sealed and may be passed through a 'shrink wrapping tunnel' for better presentation. Capacities of the unit range from five packs to 15 packs per minute.

Vacuum chamber models consist of an impulse heat sealer working within a vacuum chamber. The material to be packed is carefully placed in the laminate (heat-sealable) pouch and the open end placed on the heat sealer. Closure of the chamber grips the open end, and the chamber is evacuated to a preset residual air level. In large machines, the chamber and product can be gas flushed (usually by nitrogen gas) at this stage. This procedure extends the vacuum pack shelf-life. When the pressure inside the pouch is equal to the pressure in the chamber, the head sealer operates and closes the pack.

The chamber size determines the pack capacity that can be used. There is thus a very wide range of chamber sizes. The smallest, multi-purpose model in a table top version has a chamber volume of $0.03 \ m^3$, with an impulse seal length of 0.4 m. This machine is supplied complete with a vacuum pump, and is capable of sealing two packs (5 kg each) at approximately two per minute. Single chamber machines rise in $0.01 \ m^3$ stages up to a $0.15 \ m^3$ model which has two sealing bars of 880 mm and 450 mm. Double chamber models are available in sizes starting at $0.045 \ m^3$. Although chamber type sealing is becoming more common, the alternative evacuation and clip sealing machine is cheaper and perhaps more appropriate for small-scale operations.

All items should be packaged in cardboard cartons ready for dispatch. The use of standard pallets is recommended for the mechanised handling of large deliveries. Packaged products should be removed from the processing plant and delivered to retail outlets immediately. The product should be refrigerated as far as possible throughout the marketing chain and should not be exposed to bright light for long periods.

One last cautionary note must be made regarding packaging. Good packaging is an expensive operation which may be omitted if the product is for immediate delivery to clients after processing and if no active competition exists in the processed meat market. On the other hand, if marketing delays may not be avoided or if the product must compete against imports or other locally produced goods, good packaging may become a requirement. In this latter case, it may be worthwhile to use attractive and efficient packaging if this may result in larger sales. In all cases, the extra expense on attractive packaging must be weighed carefully against benefits derived from additional sales.

IV. LABELLING

The need for labelling is a function of the characteristics of the potential market and of the local legislation. Labelling may not be avoided if the products are to be marketed among middle- or high-income groups or are to be exported. In this case, labels play two important roles: they satisfy the need for information by literate consumers and custom officials and act as an incentive to purchase a particular product or brand. Some developing countries also require that all processed meat products be labelled. On the other hand, labelling may not be needed if goods are retailed without packaging, if the majority of the potential consumers are not literate or if no competition exists.

In case labelling is required, the following general principles should be applied:

- **The label should be legible and indelible.** Hand-written labels can be used where only a small amount of information needs to be communicated to the customer. Printed labels are advantageous in that they are clearer and allow the provision of more information. With printed labels, the layout can be planned in such a way as to emphasise important aspects of the product. Often, the consumer selects a product on the basis of name, weight and so on.

If the selling point of the product is, for instance, its value per unit weight, weight information and price could be given prominence to attract the customer's attention.

- The label should be informative. In developed countries, legislation usually dictates the minimum amount of information to be carried on the label. Developing countries are becoming more aware of labelling requirements, particularly with respect to imported food goods. Thus, most countries have enacted legislation or guide-lines which dictate what should be carried on the label. Reference to the Ministry of Agriculture of the country concerned will usually clarify the situation regarding the minimum labelling requirements. In general, these include the following categories:

- product description (e.g. fresh beef sausage): the name under which a product is sold should also include particulars as to the physical condition of the product or any specific treatment undergone (e.g. freeze-dried, smoked);

- weight: gross weight or net weight (gross weight minus wrapping or carton weight), or weight when packed if losses are likely in subsequent handling;

- ingredients: these are usually in descending order of weight, as recorded at the time of their use in the manufacture of the foodstuff;

- origin: the name or business name, and address of the manufacturer, packer or seller, and the country of origin. One may also include in this category the packing station code (if any) and date of packing; and

- durability: data on subsequent handling by the consumer (e.g. whether the product is stable at ambient temperatures or must be refrigerated) and conditions of use if these vary with time (e.g. minimum shelf-life).

- The label must be accurate. The label must not deliberately mislead the consumer. Ingredients listing in meat products is attracting great attention largely because of the difficulty of detecting adulterant meats in the product after processing. In addition, the tendency of processors to add water has led to its inclusion on the ingredients list in the European Economic Community. This presents processors with a problem since most of the water is absorbed in an uncontrolled manner, particularly in fast-curing processes. It should be noted that an ingredient which is used in a foodstuff in a dried, dehydrated or concentrated state may be placed in the ingredients list as though it had first been reconstituted. If this is the case, the ingredient should be followed by the description "when reconstituted".

- <u>Excessive labelling and detail should be avoided</u>. The Government should ensure that the prevailing legislation on labelling protects both customers and retailers and allows for fair comparison between products.

V. PROCESSING PLANT HYGIENE

A regular cleaning routine for plant, equipment and premises is essential if a high level of plant hygiene is to be guaranteed.

Wood is often a preferred surface for cutting, deboning and chopping meat. However, wooden tables or chopping blocks are very difficult to cleanse thoroughly and are not recommended. A metal table with a high-density polypropylene cutting inset affords equal protection to the edges of knives and tools and is far easier to clean.

Filling nozzles, knives, plates and so on, and the surfaces of mincers, mixers and bowl choppers require particular attention. Cleansing should involve a rough scrubbing with a brush fitted with nylon bristles to remove scraps of fat and should be followed by a thorough cleaning with a suitable bactericidal agent. Water for scrubbing down should be hot (around 82°C) and cleaning cloths should be boiled at the end of each day.

A 0.4 per cent sodium hypochlorite solution is effective for washing equipment, floors and walls. This solution should be thoroughly washed away with fresh water within ten minutes of application, to avoid corrosion of the equipment. The solution may be made up by adding 50 g of chloride of lime and 100 g of washing soda to a little cold water. The resultant paste is made up to 5 litres with water and, after settling, the clear 0.4 per cent hypochlorite solution is decanted off. The solution must be used on the day of manufacture if its powerful bactericidal properties are not to be lost. Care must be taken to ensure that the chloride of lime is not left to absorb moisture from the air. It should be kept in an airtight container.

Although cleaning schedules for plant and equipment can be arranged, it is often much more difficult to ensure that the staff comply with hygiene regulations. To reduce the risk of contamination, hot water and soap or disinfectant should be not only available but readily accessible. Education in hygiene among the workforce is essential. No outdoor clothing or footwear should be allowed in the processing area and boiler suits and aprons should be changed daily, or more frequently if they become extremely soiled. All personnel, including managers, should wear some type of head cover and should wash hands and arms thoroughly before entering the meat processing area. Smoking and eating should be prohibited in this area and all staff should be prevented from wearing jewellery or other adornment. Workers complaining of diarrhoea or stomach pains or showing signs of skin infection ord boils, should be sent home for treatment immediately.

The manager is also responsible for taking all reasonable steps in order to keep the premises clear of rats, mice, birds and insects. Direct control of flying insects during processing may be limited to using ultra-violet light traps or fly screens. The treatment of effluents is also effective in preventing the massing of flies in the roof area. The spraying of the processing plant with a suitable insecticide (e.g. Pybuthrin) each evening will go some way towards dealing with insect populations.

CHAPTER IV

PROCESSING TECHNOLOGY

Buildings and equipment for small-scale beef processing plants were described in the previous chapter for the manufacture of representative commodities from the major classes of products. Full processing details for this range of products are provided in this chapter.

The scope of the production activities is determined by a number of factors. Limited market opportunities and traditional local demand may restrict the manufacturer to the production of only one or two specific products. On the other hand, a retailer or hotelier used to imported brands may insist on the supply of a complete range of processed products. A small-scale producer should, as far as possible, be able to vary the nature of his operations accordingly.

Consideration is given to the use of muscle tissue derived from all parts of the carcase when selecting the product range. In reality, the range is more likely to reflect the nature of the raw materials that the manufacturer has available or is prepared to use. Where the object of processing is to upgrade cuts and trimmings produced by the dressing and cutting of fresh beef, the processor will be restricted to the manufacture of products such as fresh beefburger, chili con carne and perhaps charqui, which can be made from second-quality meats. A manufacturer able to use top-quality cuts, normally sold fresh, would not be so restricted in his range of products and would need to supply high value goods such as cooked, cured beef and biltong to recover manufacturing costs. Use of hindquarter and loin cuts for the manufacture of low-quality goods would be financially unjustifiable. In the following sections, first- and second-choice sources of raw materials for each of the selected products are indicated pictorially. Levels of spice additions should be adjusted, where necessary, to cater for local tastes.

The following information is provided for each meat product:

- the type of meat cuts and trimmings used in the product;
- the exact recipe for a given batch of finished product, including all spices, ice, casings and so on;
- the fractions of each ingredient per unit of output, and the yield on carcase meat "bone-in";
- a description of all processing steps, including overall processing times, and detailed processing times whenever necessary and;
- a flow diagram summarising the various steps in product preparation.

I. FRESH BEEFBURGERS

Beefburgers are preformed flat discs or balls of minced beef scraps and trimmings, sold fresh and cooked at home. The beefburgers originated from the German <u>Hackballen</u>, a finely cut meat ball, which was first manufactured more than 100 years ago. As a modern creation, it is now sold as a pattie which may be made from beef or other meat materials.

I.1 Ingredients

Beef trimmings, foreleg beef and neck are normally used for the manufacture of beefburgers, although flank and blade meat may be used where high outputs are necessary. When flank meat is used, the level of added beef fat or suet should be reduced. Gristle, bone and cartilage should be carefully removed from trimmings taken from the neck region. The meat should be chilled and lightly frozen prior to use. Beef suet from the kidney knob and channel fat should be chilled and diced into 3-4 cm cubes before being lightly frozen.

Freshness of the raw materials is the key to good-quality beefburgers. Old, bruised or dirty trimmings should not be used. Frozen trimmings are permissible providing they were of good quality and well handled prior to freezing. If hot-boned materials are used, care must be taken to ensure that contaminated matter is not carried from the slaughtering floor to the boning and processing room.

Figure IV.1 shows first- and second-choice meat cuts for the preparation of beefburgers, while table IV.1 provides estimated amounts of meat and other ingredients for the preparation of a batch of 15.64 kg of beefburgers.

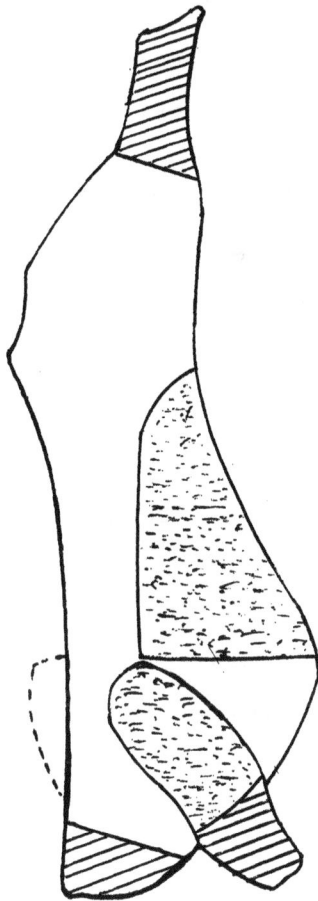

◪ First choice

▨ Second choice

Figure IV.1

Fresh beefburgers materials

Table IV.1

Ingredients for the production of beefburgers

Ingredients	Quantity of ingredients per 15.64 kg-batch	Percentage of each ingredient
Lean beef	7.34 kg	46.93
Beef trimmings (50 per cent fat)	2.86 kg	18.30
Beef fat	1.50 kg	9.60
Flake ice	3.00 kg	19.20
Powdered onion	45 g	0.29
Salt	670 g	4.28
Ground white pepper	150 g	0.96
Ground mace	75 g	0.48
Total ingredients	15.64 kg	100

Yield on finished product[1] : 100 per cent on recipe ingredients
Estimated carcase cut-out of meat inputs[2]: 82 per cent
Carcase meat "bone-in" required[3] : 11.7 kg $\div$ 0.82 = 14.270 kg
Yield on carcase meat "bone-in"[4]: (15.64 $\div$ 14.27) x 100 = <u>110 per cent</u>

I.2 Processing stages

Mincing and mixing

The pre-chilled meat and suet should be passed through a mincer fitted with an 8-mm plate and then transferred to a mixing machine. Depending on equipment available, the meat may first be chopped into 2-cm pieces with a cutter. During mincing and cutting (if applicable), it is important to keep the knives and plates as sharp as possible to minimise the crushing of the meat and to keep the temperature as low as possible. Ice and spices should be added slowly during mixing until they are fully taken up in the meat mixture. The mass should then be run through a mincer fitted with a 4-mm plate.

Forming

After final mincing, the meat should be formed into patties or balls of meat of approximately 75 g in weight. Many people prefer the meatball because the flavour and natural succulence of the meat is said to be retained if

[1] This is equal to the weight of output divided by the total weight of inputs, multiplied by 100.
[2] This is an estimate based on butchery practice.
[3] This is equal to the weight of the meat inputs (lean beef, trimmings and fat) divided by the estimated carcase cut-out of meat inputs.
[4] This is equal to the weight of output divided by the weight of carcase meat "bone-in", multiplied by 100.

cooked in this shape. For small-scale production, the flat beefburgers may be pressed out with a manually operated former. For greater production, a mechanically operated unit may be used. Alternatively, the mix may be filled into clear, regular cellulose casings with a stuffed diameter of 10 cm. After cooling to 0°C, these tubes of beefburgers meat may be cut to size on a slicing machine.

Packaging and handling

As with all fresh meat products, beefburgers should be kept under refrigeration and thoroughly cooked before being eaten. Unless the beefburgers are for immediate sale, they should be stored at -18°C throughout distribution. Fresh beefburgers are usually packaged in units of 500 g, 1, 2 or 5 kg, with a date of production clearly shown.

The complete production sequence of beefburgers is illustrated in figure IV.2.

II. CHILI CON CARNE

Chili con carne is a heavily spiced, minced beef product which is cooked almost to dryness and sold in waxed containers or casings. It is a traditional product of many South American countries, but is consumed in many other countries under different names, with minor modifications in the ingredients and processing procedures.

II.1 Ingredients

Lean beef from the shin, neck, blade and shoulder regions is normally used in the manufacture of chili con carne. Care must be taken to remove all sinews, bone and excessive fat deposits. Hot-boned beef is not recommended. Suet from the kidney or channel regions is preferred. Figure IV.3 shows the first quality materials used for the production of chili con carne, while table IV.2 provides estimates of the amounts of ingredients used for the production of a 10.84 kg batch of chili con carne.

II.2 Processing stages

Melting

The suet and onions should be passed through a mincer fitted with an 8mm-plate and placed in a steam-jacketed pan. The mixture is heated gently until the onions are light brown in colour. Overcooking should be avoided at this stage and the ingredients should not be allowed to dry out.

ESTIMATE MEAT REQUIREMENT

PRIMAL BUTCHERY

Neck, blade, trimmings

DEBONE

Lean meat   Trimmings   Fat

Weighed to give 75 per cent lean meat and 25 per cent fat

Lean meat	Trimmings	Fat
CHILL 0°C	CHILL 0°C	CHILL 0°C, 4 hrs
		DICE 3 cm chunks
		FREEZE
MINCE 8 mm	MINCE 8 mm	MINCE 8 mm

Ice, seasonings————MIX

MINCE 4 mm

FORM

PACK

DISPATCH   HOLD 0°C   FREEZE -18°C

DISPATCH 0°C

HOLD 2 months

DISPATCH -18°C

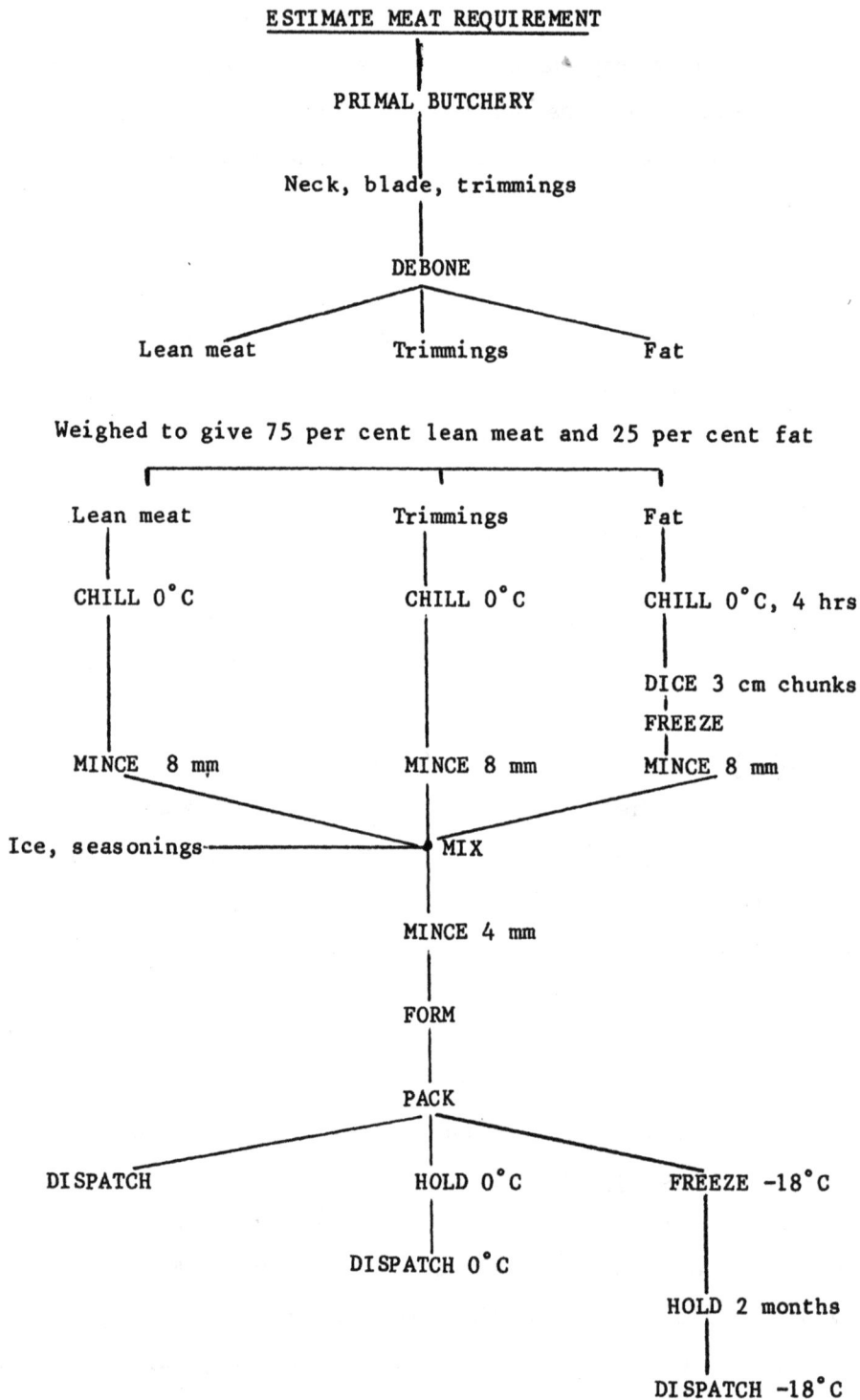

Processing time : 8 hrs

Yield on 'bone-in' meat : 110 per cent

Figure IV.2

Flow diagram illustrating beefburger preparation

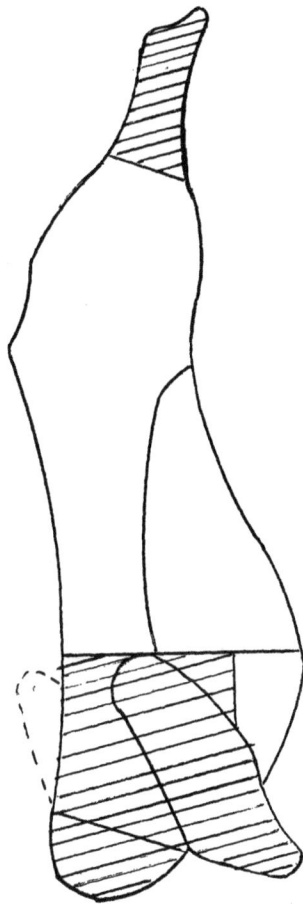

First quality

Figure IV.3

Chili con carne materials

Table IV.2

Ingredients for chili con carne

Ingredients	Quantity of ingredients per 10.84 kg batch	Percentage of each ingredient
Lean beef	7.0 kg	61.32
Kidney suet	2.2 kg	19.27
Cassava flour	800 g	7.00
Salt	250 g	2.19
Chili pepper	400 g	3.50
Paprika	150 g	1.31
Coriander	35 g	
Ground cumin seed	50 g	
Ground oregano	20 g	1.00
Ground nutmeg	10 g	
Fresh,chopped onion	500 g	4.38
Total ingredients	11.415 kg	100

Yield on finished product: 95 per cent on recipe ingredients

Estimated carcase cut-out of meat ingredients: 84 per cent

Carcase meat "bone-in" required: 9.200 kg ÷ 0.84 = 10.952 kg

Yield on carcase meat "bone-in": [(11.415 x 0.95) ÷ 10.952] x 100 = 99 per cent.

Cooking

The beef should be passed through a mincer fitted with an 8-mm plate, and the minced meat added to the melted fat and onions. The salt should also be added at this stage. Cooking is continued until no further steam is driven off. Constant stirring is necessary to avoid scorching the product. The mass should not be allowed to harden. Spices and sifted flour should be added during cooking, preferably 15 minutes before completion. The mass must be thoroughly stirred during this addition to prevent it becoming lumpy.

Filling and handling

After the mass has been thoroughly cooked, it should be filled into rectangular moulds and chilled overnight. The size of the mould will depend upon local market requirements. The following day, the mould should be dipped into hot water for 15 seconds to melt the fat surface in contact with the mould. The product can then be removed by inverting the mould on to greaseproof paper.

Packing and storage

The product should be wrapped either in greaseproof paper or sealed in plastic bags. An alternative procedure is to fill the chili con carne into cellulose casings or beef middles and tie off into 20-cm lengths. The wrapped product must be stored under refrigeration and should be consumed within five days of manufacture.

The complete production sequence is illustrated in figure IV.4.

III. BILTONG

The art of producing air-dried beef products has evolved over a considerable period of time in view of the need to preserve meat surpluses associated with seasonal distortions in the level of slaughter. In recent times, other preservation methods such as chilling and freezing have assumed a greater importance, but dried beef is still consumed in many areas of the world. The main dried beef products are biltong, charqui and pemmican. Similar products, known under various other names, may be found in different countries. Meat can be dried directly, as in the case of charqui or pemmican, or after prior treatment to improve the quality of the finished product, as in the case of biltong.

The major difference between biltong and other dried beef products is that it is not cooked either during processing or prior to consumption. Biltong is cured and then dried so that approximately 60 per cent of the mass of the meat is lost. The period of drying and the manner in which it is carried out is flexible and a matter for the processor's own judgement.

III.1 Ingredients

Biltong is generally made from hindquarter muscles of young, lean cattle. These are dissected along their seams and cut into strips 25-30 cm long, 5-10 cm wide and 2-4 cm thick. When the climate is very hot and dry, thicker strips (5-7 cm)) may be used. A comparatively lean hinquarter will yield about 70 per cent meat for biltong, 12 per cent trimmings and 18 per cent bone.

ESTIMATE MEAT REQUIREMENT   (ESTIMATE CASING REQUIREMENT)

PRIMAL BUTCHERY   Beef middles

Shoulder, shin, neck, kidney knob, channel fat

DEBONE

Lean meat   Trimmings   Fat

Weighed to give 75% lean, 25% fat   SOAK 20°C, 2 hrs

Lean + trimmings   Fat, onions
MINCE 8 mm   MINCE 8 mm
                            HEAT

COOK 50%

────── Add spices, flour

COOK 50% ──────────────────────────────── COOL 15°C

FILL MOULD   FILL CASINGS

CHILL 0°C, 24 hrs   CHILL 0°C, 24 hrs

HOT WATER DIP

REMOVE MOULD

PACK (greaseproof paper)

DISPATCH 0°C   DISPATCH 0°C

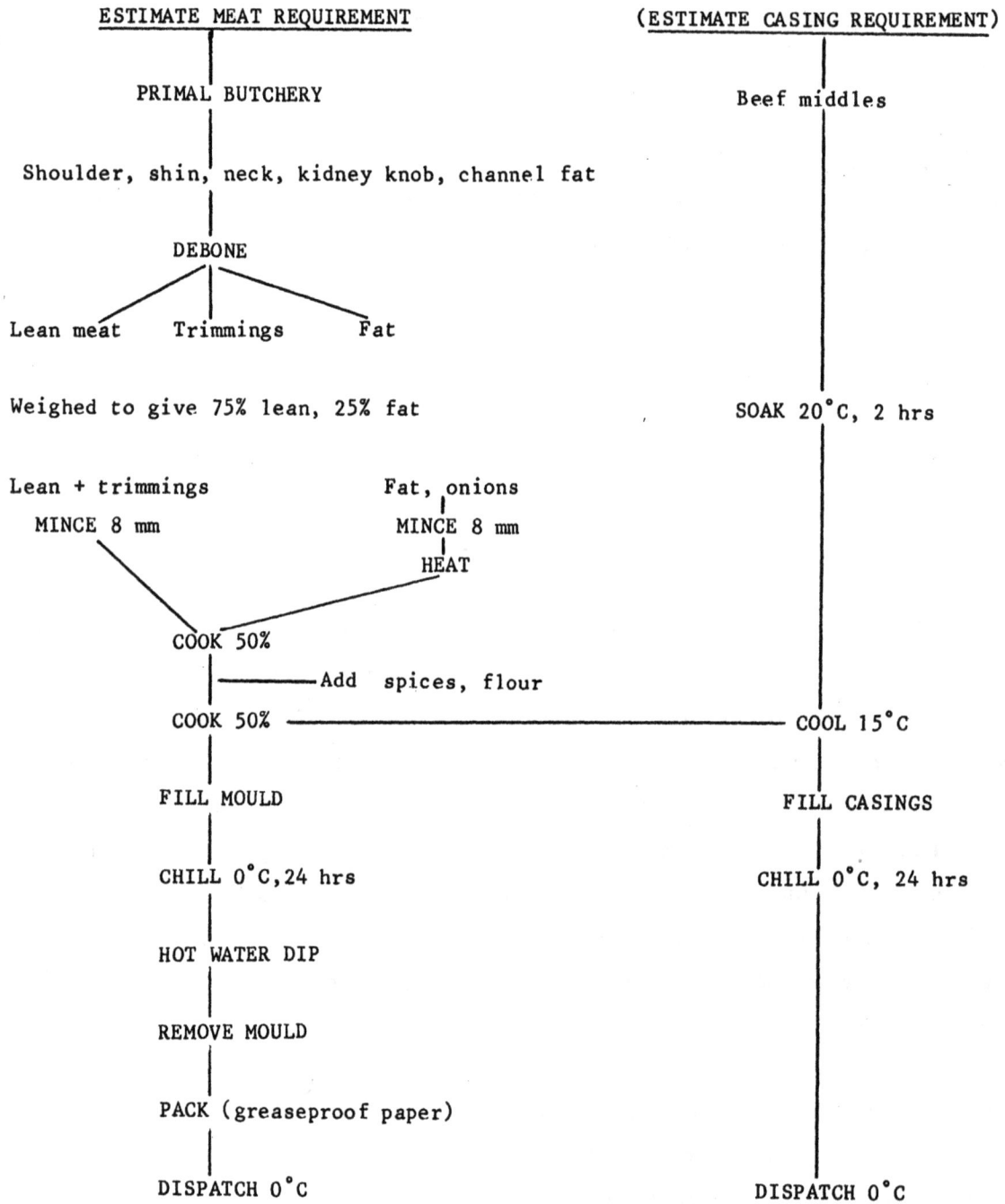

Processing time: 8 hrs
Yield on "bone-in" meat: 99 per cent

Figure IV.4

Flow diagram illustrating chili con carne preparation

Strips of muscle cut from unchilled carcases within one to five hours of death are most suitable for the manufacture of biltong. Chilled muscle may also be used if hot-boned material is unavailable. Only carcases from approved slaughterhouses which have been subjected to veterinary inspection should be used.

Figure IV.5 shows first- and second-choice meat cuts for the production of biltong, while table IV.3 provides estimates of the amounts of ingredients needed for the production of a 10-kg batch of product.

Table IV.3

Ingredients for the production of biltong

Ingredients	Quantity of ingredients per 10-kg batch of finished product
Lean beef	23.750 kg
Immersion brine	
Drinking water	42 litres
Nitrite salt	5.740 kg
Brown sugar	73 g
Ground coriander	7 g
Ground black pepper	7 g

Weight of finished product : 10 kg

Yield on finished product: approximately 40 per cent on recipe ingredients (this allows for salt and spice uptake)

Estimated carcase cut-out of meat inputs: 82 per cent

Carcase meat "bone-in" required: 23.750 kg ÷ 0.82 = 28.960 kg

Yield on carcase meat 'bone-in' : (10 ÷ 28.960) x 100 = 35 per cent

III.2 Processing stages

Curing

Biltong may be salted either by packing in dry salt or by immersion in brine. The immersion technique is favoured.

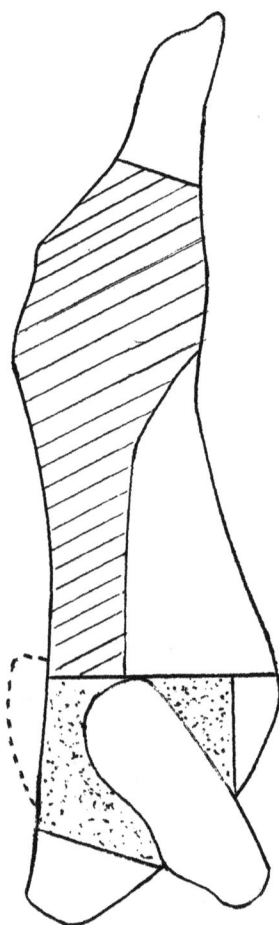

First choice
Second choice

Figure IV.5

Biltong materials

After cutting the hindquarters and loin muscles into strips, these should be placed in a tank (plastic, food grade enamel or stainless steel vessels are suitable) and completely covered in brine. A weight may be used to ensure complete submersion. A ratio of one part by weight of meat to two parts by weight of brine should be aimed for. The length of the curing period will depend upon the leanness of the meat used and the degree of subsequent drying, as well as personal taste. Biltong which contains a lot of fat will absorb salt more slowly than lean biltong. Furthermore, the finished product will become more salty the longer it is left to dehydrate. The finished salt content of the product is dependent on local tastes. Generally, for pieces of meat 4-cm thick, 36-40 hours should be a sufficient immersion time. Larger pieces (e.g. 7-cm thick) should be left in brine for three to four days. The temperature of the brine should not be allowed to exceed 7°C; this may be achieved by storage in a refrigerated room.

Upon removal from the tank, the strips of meat may be dipped for several minutes in 1 per cent acetic or lactic acid solution to reduce the risk of microbial spoilage during drying.

Drying

Drying is carried out by hanging the salted meat strips on wire hooks suspended from strands of galvanised wire. It is important that the strips do not touch and that drying is carried out in a controlled environment, under microbiologically clean conditions. The meat should be hung under cover in a shaded, airy place. It should be protected against flies, insects, dust and rain. Protection against flying insects can be achieved physically by netting the drying enclosure. Crawling insects may be eliminated by regular insecticide application to the area surrounding the drying grounds. The latter should have a concrete base.

In areas of moderate humidity, an artificial drying chamber might be necessary to initiate the drying cycle. Air dried goods cannot be produced successfully in regions of high humidity (e.g. relative humidity of 90 per cent). Thus, the production of other products should be contemplated.

After one day's drying, the strips should be taken off the wire, straightened out and re-suspended with the opposite end uppermost. The time taken for drying will depend on the weather. The biltong is considered to be ready for storage or consumption when a piece, broken or cut off, shows a uniform structure. This is equivalent to a weight loss after curing of approximately 60 per cent. Thus, a 5-kg cut of hindquarter muscle should yield approximately 2 kg of biltong.

Storage and packaging

Although fat may become rancid, biltong should keep for six months or more if stored in a dried place away from insect attack. As such, it can be wrapped in a variety of materials such as cheesecloth or palm fronds, depending on local availability. It may also be vacuum packed or kept in frozen storage where it may be expected to keep indefinitely.

The complete processing procedure is illustrated in figure IV.6.

IV. CHARQUI

Charqui, which is produced in vast quantities in Latin America, differs from biltong in that it is a very fat product. It is produced from the belly and flank region of the carcase. Additionally, it is not treated prior to drying and is cooked before being eaten.

IV.1 Ingredients

In many Latin Amerian countries, beef sides are divided according to the "Pistola" cutting procedure into forequarters, special hindquarter and flank. The flank (or "ponta de agulha" piece), is removed by freeing the muscles of the abdomen from the proximal pelvic limb at a distance of 24 cm from the back-bone and then cutting between the fifth and sixth ribs. The rib bones must be sawn through and then carefully removed to avoid damage to the underlying musculature. The septa covering the inside of the cut should also be removed. The best grade charqui contains 20-35 per cent of fatty tissue.

Figure IV.7 shows the part of the carcase used for the production of first-choice charqui.

IV.2 Processing stages

Salting

The deboned and trimmed whole flank pieces should be stacked into heaps 1-1.5 metres high on a slightly sloping concrete floor. Coarse, dry salt is

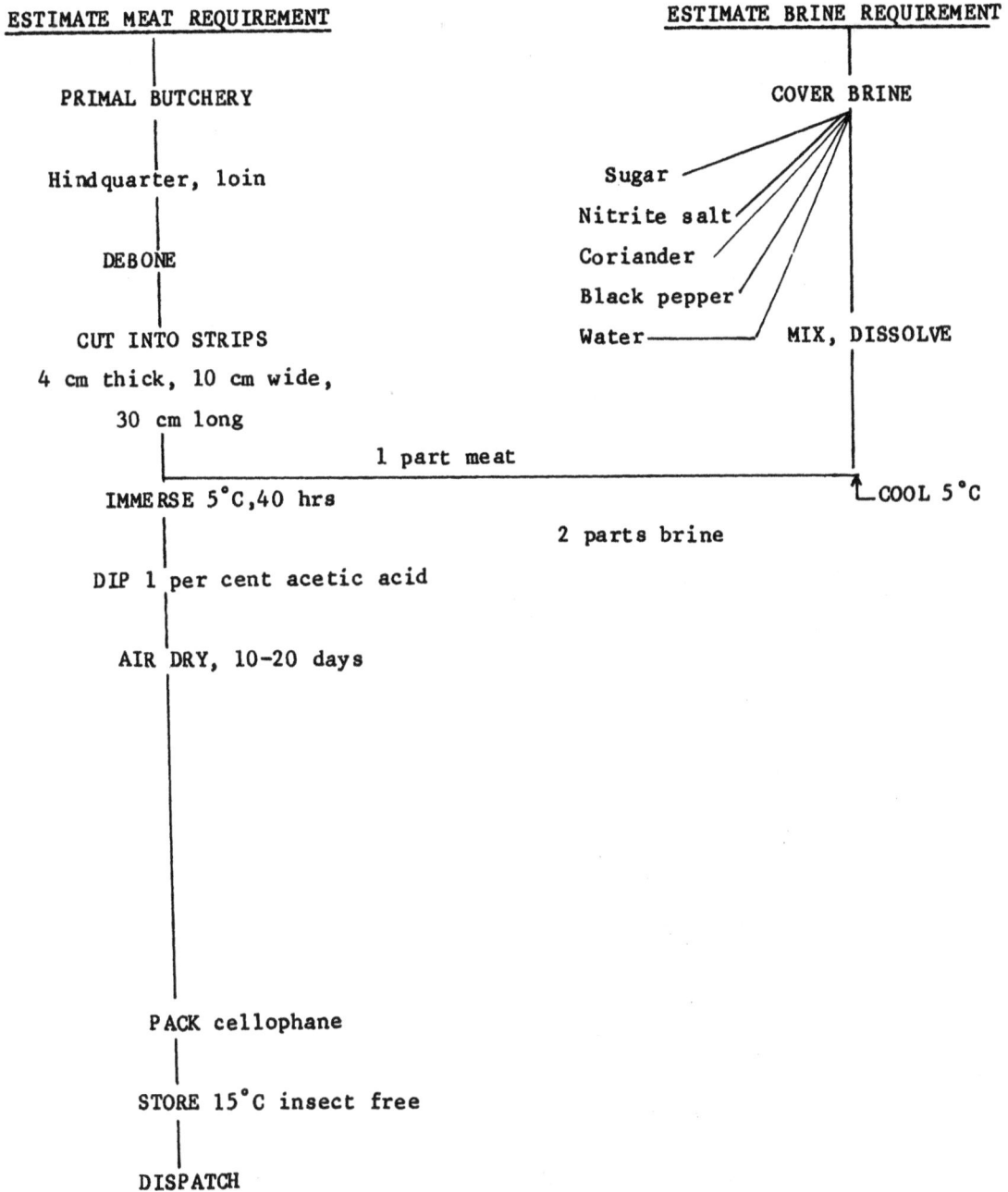

ESTIMATE MEAT REQUIREMENT   ESTIMATE BRINE REQUIREMENT

PRIMAL BUTCHERY   COVER BRINE

Hindquarter, loin   Sugar

Nitrite salt

DEBONE   Coriander

Black pepper

CUT INTO STRIPS   Water   MIX, DISSOLVE

4 cm thick, 10 cm wide,

30 cm long

1 part meat

IMMERSE 5°C, 40 hrs   COOL 5°C

2 parts brine

DIP 1 per cent acetic acid

AIR DRY, 10-20 days

PACK cellophane

STORE 15°C insect free

DISPATCH

Processing time: 300-500 hrs

Yield on 'bone-in' meat: 35 per cent

Figure IV.6

Flow diagram illustrating biltong preparation

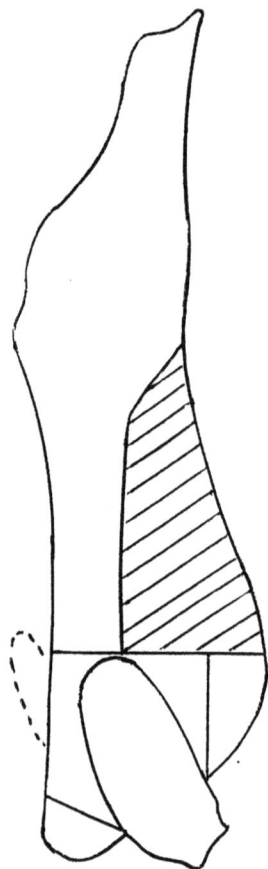

First choice

Figure IV.7

Charqui materials

liberally scattered between each layer and on the top of the pile which is left overnight. The piles are turned each day for four days so that the pieces from the top of the pile go to the bottom of the new pile and those from the bottom go to the top. The piles should be covered with fresh salt each day and their temperature should not be allowed to exceed 20°C. The coarse salt dissolves and recrystallises on the surface of the product to give a fine salt layer. Any undissolved particles of coarse salt should be brushed off and discarded.

Drying and curing

Drying begins on the fifth day. The meat should be hung over wooden drying racks and exposed to the sun for no more than one to two hours. It should then be removed from the racks, resalted, piled into stacks about 1 metre high under a tarpaulin and left to 'cure' for two to three days. Drying and curing is repeated five to seven times until the meat has lost 40 to 60 per cent of its fresh weight (depending on fat content). Drying must be carefully carried out to avoid overheating the meat and causing the fat to melt. The exposure time may be increased as the product dries (e.g. from two hours to six hours). The salt on the surface of the dried meat should be fine and perfectly white, and a final salting should be made with fine salt to improve the appearance of the product.

Storage and packing

The production of charqui is a slow, labour-intensive process. It is open to criticism on hygienic grounds. Nevertheless, if the processing is carefully controlled, charqui may be stored for up to two months without loss of quality. Charqui must also be vacuum packed in plastic film if the shelf-life of the product is to be further extended.

The complete production sequence is illustrated in figure IV.8.

IV.3 Processing yield

The yield (finished product) on carcase meat 'bone-in' is difficult to estimate because it depends on the leanness of the raw material, the duration of the drying and curing operations, and so on. The charqui recipe which is used in the costing example in Chapter V is based on a yield of 24 per cent of carcase meat "bone-in". This is estimated as follows:

Weight of finished product = 10 kg

Meat (flank)   25 kg

Salt   2.250 kg

Approximate yield on processing ingredients: 38 per cent

- 64 -

ESTIMATE MEAT REQUIREMENT
PRIMAL BUTCHERY

Flank

DEBONE

SALT

STACK 20°C   Turn the
                              stack and resalt each day
TURN STACK 20°C   for four consecutive days

RESALT

TURN STACK

DRY 1-2 hrs

RESALT

Repeat   STACK   (curing for 2 to 3 days)
5 to 7 times

DRY 3 hrs

Etc.

DRY 6 hrs

40-60 per cent yield on meat

PACK

Cellophane,   Vacuum
Cheesecloth, local   STORE, ambient, 3 months
HOLD 1-8 weeks   DISPATCH   DISPATCH

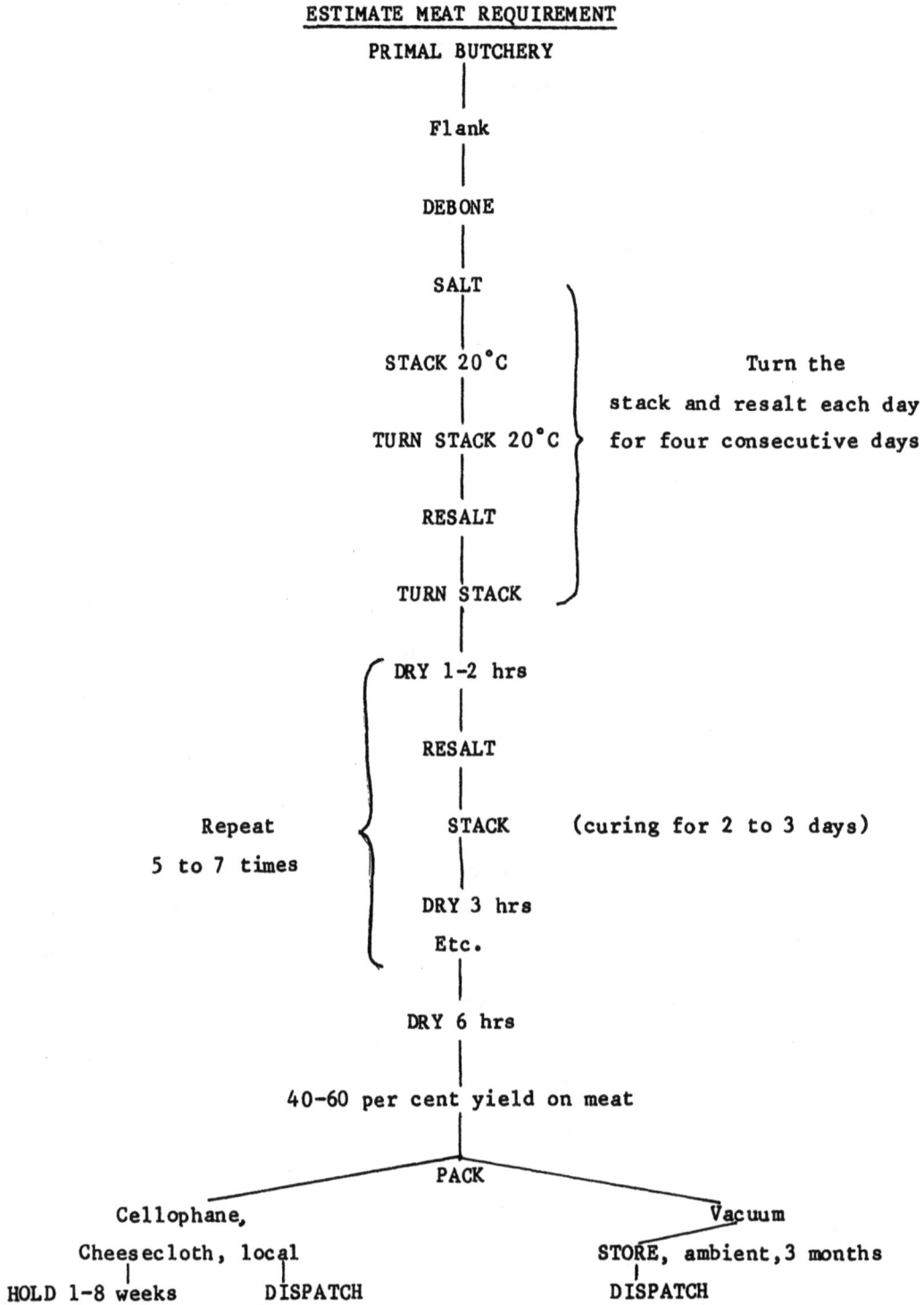

Processing time: 21 days
Yield on "bone-in" meat: 25-40 per cent

Figure IV.8
Flow diagram illustrating charqui preparation

Estimated carcase cut-out including trimming losses: 60 per cent

Carcase meat 'bone-in' required= 25 kg ÷ 0.6 = 42 kg

Yield on carcase meat 'bone-in' = (10÷42) x 100 = <u>24 per cent</u>

With other raw materials, the yield on carcase meat "bone-in" could be as high as 40 per cent.

V. COOKED, CURED BEEF

Cooked, cured beef products may be prepared from any muscle from the hindquarter. However, the silverside and brisket cuts are the most commonly used. The products are cured in such a way as to produce a distinctive red colour. They are then cooked in moulds. Cured silverside may be smoked before cooking. Cooked, cured beef products are usually sold sliced and may be used as fillings for sandwiches or for salads.

V.1 Ingredients

For the preparation of cured silverside and cured rolled brisket, it is necessary to use muscles from young animals or those without excessive levels of fat.

The silverside (<u>Biceps femoris</u> and <u>Semitendinosus</u> muscles) is removed from the hindquarter and divided into three blocks, two from the <u>Semitendinosus</u> muscle so that the fibre direction runs parallel to the long axis of the block. Brisket (<u>Pectorales profundi</u>, <u>Serratus ventralis</u>) is removed from the forequarter by trimming away from the rib cage. The muscle is rolled so that the majority of muscle fibres are parallel to the longitudinal axis and tied up at 10-cm intervals. This cut tapers to a thin muscle. A more even roll is obtained by turning the taper in on itself. The rolls are chilled to 2°C and trimmed of surface fat to a 4-mm depth.

Figure IV.9 shows first-choice meat cut for the production of cooked, cured beef while table IV.4 provides estimates of the amounts of ingredients needed to produce a 10-kg batch of product.

V.2 Processing stages

Curing

The muscles should be injected with <u>pumping brine</u> to the point where the fresh meat weight is increased by 8 to 10 per cent. A hollow needle with a number of holes along its length may be used to force the brine directly into

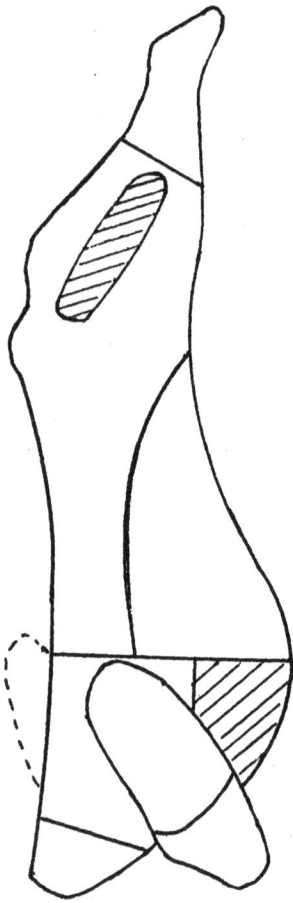

Figure IV.9

Cooked, cured beef materials

//// First choice

Figure IV.10

A cured meat mould

Table IV.4

Ingredients for the production of cooked, cured beef

Ingredients	Quantity of ingredients per 10-kg batch of finished product
Lean meat	10.7 kg
Pumping brine	
Nitrite salt	130 g
Sugar	2.5 g
Whole black pepper	
Bay leaves	
Dried mace	To taste
Marjoram	
Water	1.0 litre
Immersion brine	
Nitrite salt	2.680 kg
Sugar	52 g
Whole black pepper	
Bay leaves	To taste
Dried mace	
Marjoram	
Water	20 litres

Weight of finished product: 10 kg

Yield on finished product: 94 per cent on meat ingredients

(this allows for salt and spice uptake during pumping, etc.)

Estimated carcase cut-out of meat input: 82 per cent

Carcase meat "bone-in" required: 10.7 kg $\div$ 0.82 = 13 kg

Yield on carcase meat 'bone-in' = (10 $\div$ 13) x 100 = 77 per cent the muscle. Several sites should be chosen to ensure uniform distribution of curing salts. After injection of the brine, the muscle pieces should be placed in a curing tank and covered with immersion brine at the same concentration as that used for the pumping brine. A ratio of one part of meat to two parts by weight of brine should be used. A weight may be placed on the uppermost piece to ensure complete submersion. The muscles should be left in the brine for two days at a temperature which must not exceed 7°C. Since a reduction in brine strength and bacterial contamination are inevitable, the brine should not be reused.

Smoking (where applicable)

Smoking is often carried out on beef silverside in order to produce a characteristic flavour. It is generally not recommended for the rolled brisket. The muscle should be suspended in the smokehouse for one to two hours prior to smoking in order to promote surface drying. A smoking period of 20 hours at 24°C should be sufficient to produce the desired golden yellow colour.

Cooking

After curing (and eventually smoking), the cured muscles should be weighed and then pressed firmly into a rectangular mould. Slight pressure should be applied to the retaining, spring-loaded lid in order to guarantee a satisfactory product shape. Since cooking procedures for the two products are slightly different, they are described separately in the following sections.

Cooking of silverside

Cooking takes place in a meat cooker. The rectangular silverside moulds (see figure IV.10) should be immersed in water at 100°C for 20 minutes in order to seal the surface of the product and to lessen cooking losses. Thereafter, they should be cooked in water at 78°C for 50 minutes per kg of meat. A 2-kg product would therefore be cooked for 20 minutes at 100°C followed by 100 minutes at 78°C (i.e. a total cooking time of two hours).

Cooking of rolled brisket

Cooking takes place in a meat cooker. The brisket moulds should be immersed in 78°C water for one hour per kg of meat. A 3-kg product will thus be cooked for three hours at 78°C.

After cooking has been completed, the moulds should be placed in a cold room at 2°C. The products should be removed from the forms on the following day.

Packaging and storage

All products should be stored under refrigeration. When vacuum packaged, rolled cured brisket and cured silverside will have a shelf-life of 14 to 21 days at 0°C. If wrapped in a plastic film and held at 0°C, a shelf-life of five to 10 days may be anticipated.

The complete production sequence of cooked, cured beef is illustrated in figure IV.11.

VI. FRANKFURTERS

Frankfurters are one of the most popular of all sausage products. They may be made from all beef or various combinations of meats. The manufacture of frankfurters is typical of the production of most fine-cut sausages stuffed into narrow casings. They are encased, linked, smoked and cooked and may be sold skinless or in their casings, depending upon local taste.

VI.1 Ingredients

Frankfurters are made from fresh, uncured meat. For the production of high quality, all-beef products it is normal to use only meat cuts from the shoulder and leg regions. Trimmed meat from the neck may also be included. The lean beef, beef suet, dripping and beef trimmings should be chilled prior to processing.

Figure IV.12 shows first- and second-choice meat cuts used in the production of high- and low-quality frankfurters, while table IV.5 provides estimates of the amounts of ingredients required for the production of batches of high- and low-quality frankfurters respectively.

VI.2 Processing stages

Mincing

The lean meat and fat should be minced separately through a 4-mm plate.

Chopping

After the meat has been minced, it should be placed in a bowl cutter with the nitrite salt, seasonings and one-quarter of the ice. The whole mixture shoud then be chopped for several revolutions. It is necessary to restrict

ESTIMATE MEAT REQUIREMENT   ESTIMATE BRINE REQUIREMENT

PRIMAL BUTCHERY   PUMPING BRINE   COVER BRINE

Silverside, Brisket

Sugar
Nitrite Salt
Black Pepper
Bay Leaves
Mace
Marjoram
Water

DEBONE

ROLL AND TRUSS MIX DISSOLVE   MIX DISSOLVE

PUMPING BRINE (110%) ———————— COOL 5^0C

1 part meat   COOL 5^0C

IMMERSE 7^0C, 2 days ——————— 2 parts brine ————

DRAIN

Brisket   Silverside

SMOKE (20 hrs, 24^0C)

MOULD   MOULD

COOK 78^0C, 60 mn/kg COOK 100^0C, 20 mins
        78^0C, 50 min/kg

COOL 2^0C, 24 hrs

REMOVE MOULD

PACK

Cellophane   Vacuum ————— HOLD 0^0, 10 days

DISPATCH, 5^0C DISPATCH 0^0C DISPATCH, 0^0C

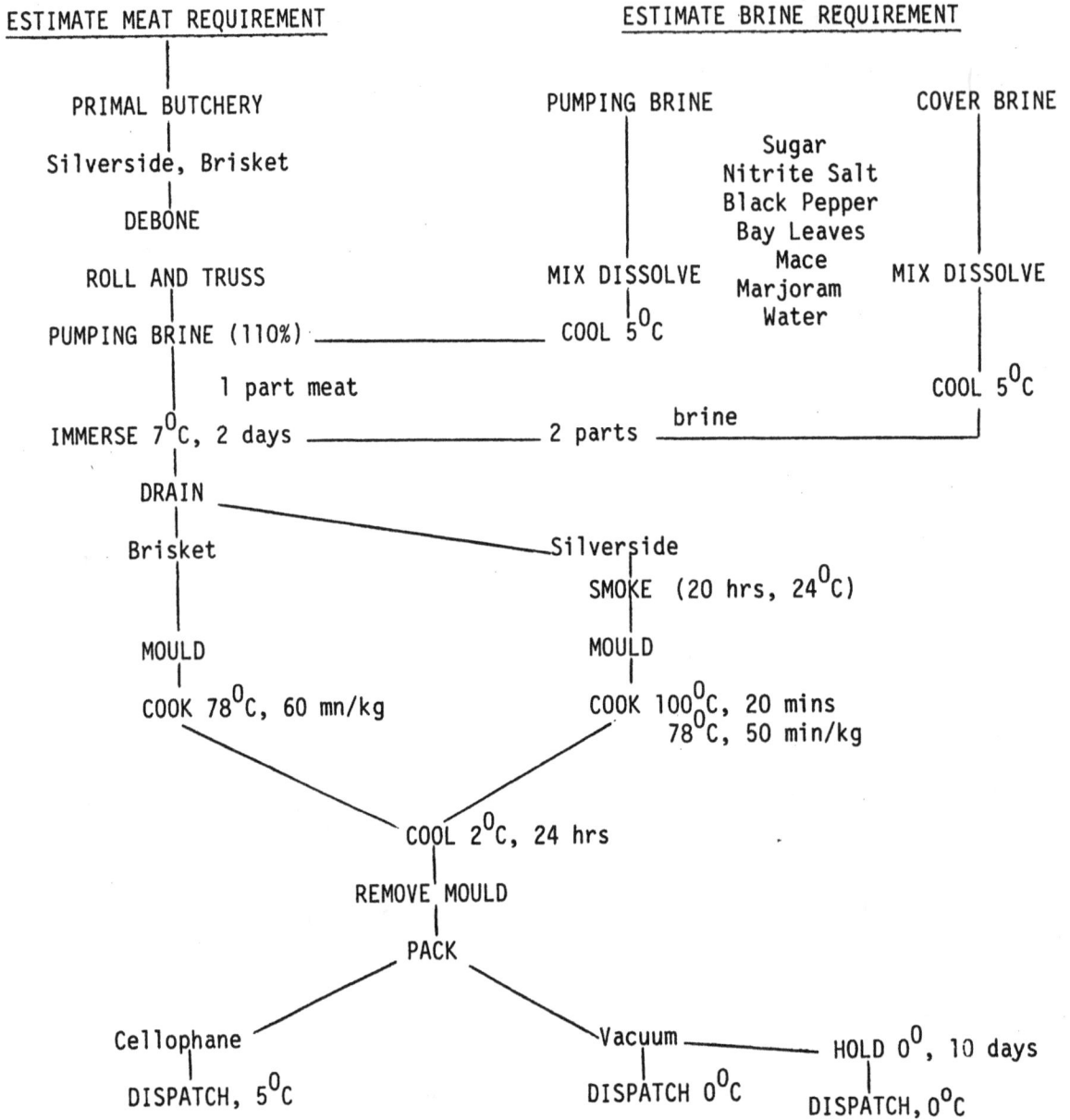

Processing time: 3 days

Yield on "bone-in" meat: 77 per cent

Figure IV.11

Flow diagram illustrating cooked, cured beef preparation

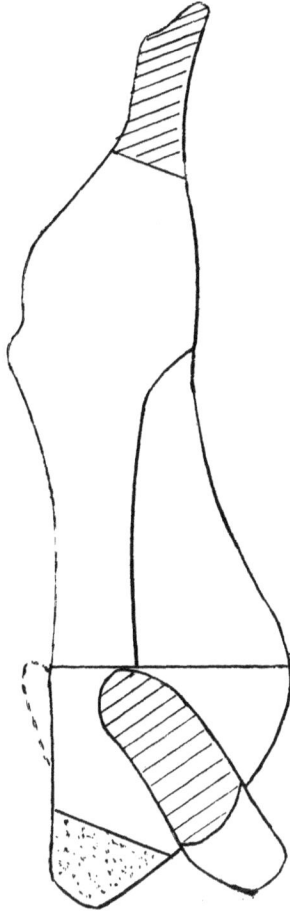

/// First choice

::: Second choice

Figure IV.12

Frankfurter materials

ice addition at this stage in order to ensure a maximum extraction of the salt soluble contractile proteins which improve the binding of the finished product. Finally, the fat and the remainder of the ice are added and chopped until the temperature reaches 15°C. Temperature, rather than time, is a better guide in chopping. Other ingredients, such as non-meat proteins (cereals and milk powders), are added during the final ice addition since they will readily absorb water and interfere with the extraction of soluble meat protein.

Table IV.5
Ingredients for the production of frankfurters

Ingredients	Quantity of ingredients per batch of finished product	Percentage of each ingredient
A. High quality frankfurters (20.77-kg batch)		
Lean beef	14.0 kg	64.00
Beef suet	4.0 kg	18.30
Beef dripping	2.0 kg	9.15
Ice	1.4 kg	6.40
Nitrite salt	400 g	1.83
Ground white pepper	40 g	
Ground mace	20 g	
Ground ginger	4 g	0.32
Ground coriander	4 g	
Powdered lemon peel	2 g	
Total ingredients	21.870 kg	100.00

Yield on finished product: 95 per cent of recipe ingredients
Estimated carcase cut-out of meat inputs: 84 per cent
Carcase meat "bone-in" required : 18 kg ÷ 0.84 = 21.43 kg
Yield on carcase meat 'bone-in': $((21.870 \times 0.95) \div 21.43) \times 100$ = 97 per cent

B. Low quality frankfurters (25.79-kg batch)		
Lean beef	9.0 kg	32.80
Regular beef trimmings (70 per cent fat)	7.0 kg	25.51
Beef suet	4.0 kg	14.57
Ice	5.9 kg	21.50
Dry skim milk powder	886 g	3.23
Nitrite salt	590 g	2.15
Ground white pepper	47 g	
Ground nutmeg	12 g	0.24
Garlic powder	3 g	
Total ingredients	27.438 kg	100.00

Yield on finished product: 94 per cent on recipe ingredients
Estimated carcase cut-out of meat inputs: 80 per cent
Carcase meat "bone-in" required : 20 kg ÷ 0.8 = 25 kg
Yield on carcase meat "bone-in" = $((27.438 \times 0.94) \div 25) \times 100$ = 103 per cent

Stuffing

Small cellulose or sheep casings are generally used in the manufacture of frankfurters. Cellulose casings ensure that the sausages will have a uniform diameter and length. They also better withstand modern, high-temperature, rapid processing conditions. They are easy to stuff, possess a high degree of resilience to breakage and are permeable to smoke when moist. Small cellulose casings are manufactured in lengths of 20 to 50 metres and stuffed diameters of 15 to 30 mm.

Sheep casings are usually sold in 100 metre hanks and graded according to stuffed diameter. The latter may range between 16 and 27 mm. Depending on grade, 10 kg of chopped meat will fill between 33 and 60 metres of casings. Sheep casings should be flushed with water prior to being placed on the stuffing horn. Excess water is removed during stuffing by stripping between the fingers. As the casing fills, it should be held firmly between the fingers to eliminate air pockets.

Linking

As the casing fills, it is allowed to work off the stuffing horn on to a table. The tube of sausage meat may be divided into lengths of 12 to 13 cm either by manual linking or by automatic linking machines. Although machines are time-saving, they are expensive and can only be justified for capacities exceeding 250 kg/hour.

Smoking and cooking

After linking, the sausage is hung on smoking sticks and sprayed with cold water to wash off adhering meat. The links are separated on the sticks in order to ensure a free movement of smoke between the frankfurters. Smoking times and temperatures vary considerably according to plant facilities. A simple smoking and cooking cycle lasts for 2.5-3 hours, during which time the centre sausage temperature is raised to 68°C. Smoke is applied during the first 30 minutes in order to produce a desirable surface colour and to develop the surface skin. The temperature of the smokehouse should be held at 60°C and the relative humidity at 40 per cent during the smoking stage. After smoking, the temperature is raised to 80°C for final cooking.

When animal casings are used, the initial smoking step is preceded by the drying of the sausage surface. This is important in providing a surface to which smoke will adhere rather than penetrate.

Chilling

Franfurters should be chilled immediately after cooking to an internal temperature of 30°C. This causes the sausage to swell slightly and improves its appearance. Spraying with cold water is particularly effective. After chilling, the sausages should be held in a cold store to cool further and dry. Excessive removal of moisture from the product should be avoided.

Packaging and storage

Frankfurters may be stored and sold unwrapped provided they are chilled. The shelf-life of the unwrapped, fresh product is no longer than 10 days, after which the skin will shrivel and the flavour will become rancid. If frankfurters are vacuum-packed, they could last for up to five weeks.

The complete production sequence of frankfurters is illustrated in figure IV.13.

VII. BEEF CERVELAT

Beef cervelat is a cured, coarse cut, semi-dry sausage which is smoked but not cooked before dispatch. Like the majority of sausage products, beef cervelat may be produced from a variety of meats or meat combinations.

VII.1 Ingredients

Meat cuts from the shoulder, neck and flank regions can be included. All cuts should have a dry texture and no trace of blood on the surface (cuts from the neck in particular). The meat should be chilled (7°C) prior to processing.

Figure IV.14 shows first- and second-choice cuts for the production of beef cervelat while table IV.6 provides estimates of the amounts of ingredients required for the production of 17.77-kg of product.

VII.2 Processing stages

Mincing

The lean meat is minced through a 6-mm plate, while the trimmings are minced through an 8-mm plate.

ESTIMATE MEAT REQUIREMENTS   ESTIMATE CASING REQUIREMENT

PRIMAL BUTCHERY   CASINGS
(Sheep narrow
or cellulose)

Shoulder, neck, lean trimmings

DEBONE

Lean meat   Fat   Back fat

Weighted to give 55% lean meat
45% fat

Lean meat   Fat

MINCE 4 mm   MINCE 4 mm

25% ice
nitrite salt ———→ BOWL CUTTER (To chop 5 revolutions)   SOAK 25°C, 2 hrs
seasonings

75% ice ———————→ BOWL CUTTER (To chop up to 15°C)←—
other ingredients

Prepare stuffer——————————————————

STUFF (4 m per kilogram)

LINK (12-13 cm long)

SMOKE 60°C, 30 mn

COOK 80°C, (internal temperature)

SPRAY COOL 30°C (internal temperature)

DRY 0°C, 24 hrs

PACK

Cellophane   Vacuum pack

DISPATCH 0°C   STORE 0°C, 1-6 weeks
DISPATCH 0°C

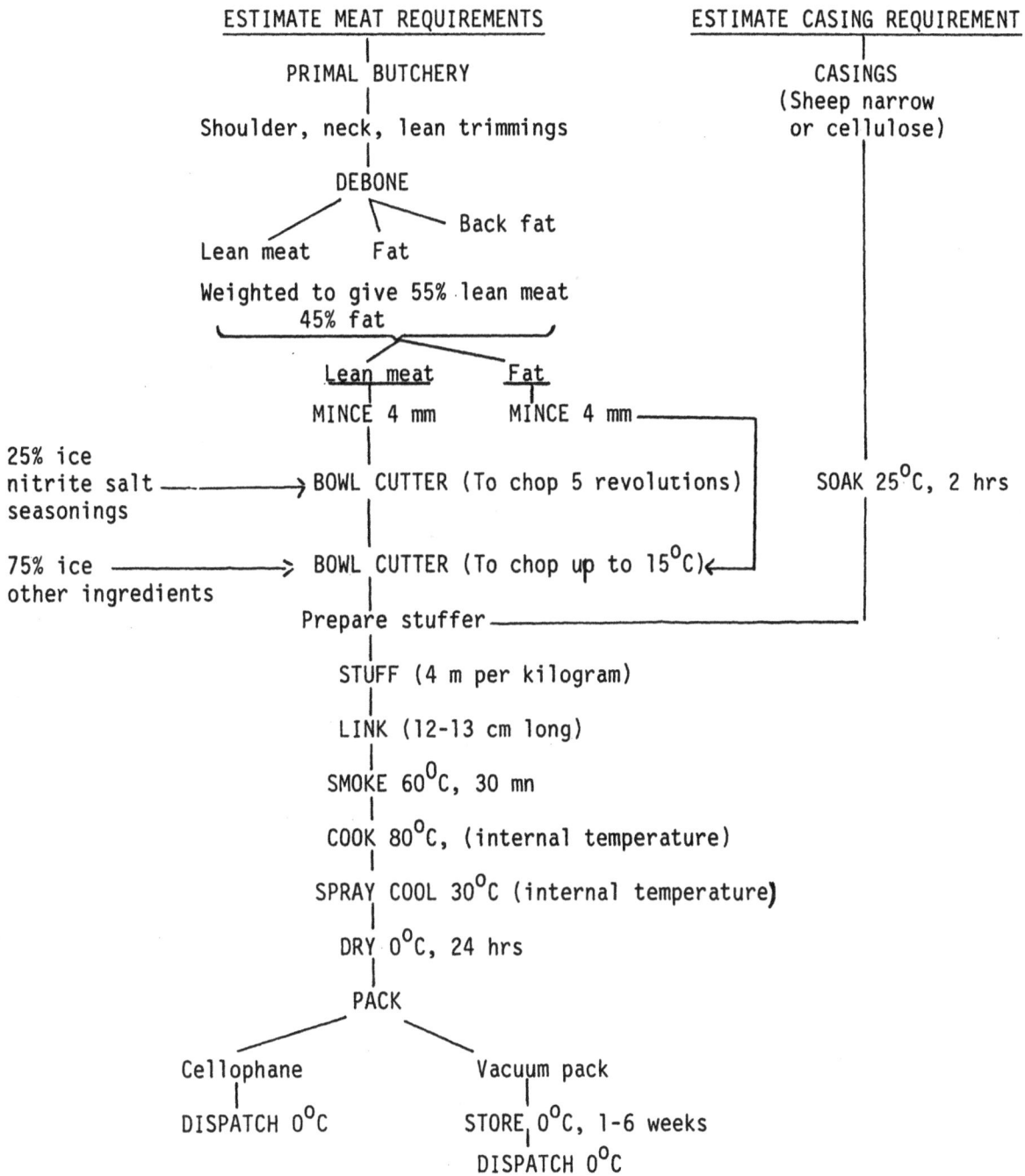

Processing time: 48 hrs

Yield on "bone-in" meat: 97 per cent for high-quality frankfurters
103 per cent for low-quality frankfurters

Figure IV.13

Flow diagram illustrating frankfurter preparation

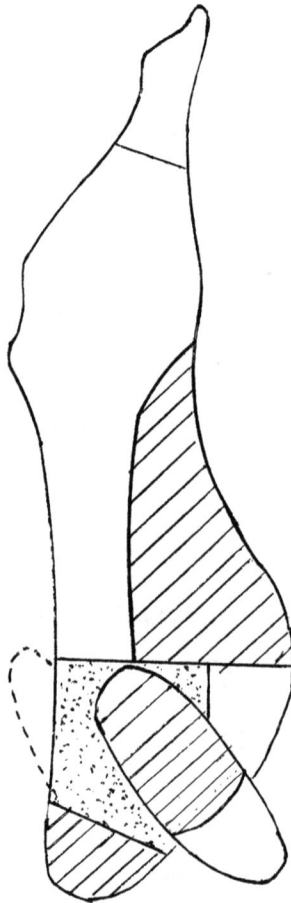

/// First choice
Second choice

Figure IV.14

Beef cervelat material

Mixing

The minced meat is mixed with all the remaining ingredients, except the whole black pepper. The mixture is re-minced through a 3 mm plate, and re-mixed with the whole black pepper in a mixer fitted with a dough hook. This kneading-mixing should be carried out for 2-5 minutes. The mixture is then transferred to a deep pan and held at 5°C for 72 hours to cure.

Stuffing

Prior to stuffing, the sausage is re-mixed. It is then stuffed into either beef middles or No. 2 size fibrous casings and tied off in lengths of 35 cm. Air pockets should be carefully avoided.

Table IV.6
Ingredients for the production of beef cervelat

Ingredients	Quantity of ingredients per 17.77-kg batch of finished product	Percentage of each ingredient
Lean beef	14 kg	66.98
Beef trimmings (50 per cent fat)	6 kg	28.70
Salt	600 g	2.87
Sugar	200 g	0.96
Ground black pepper	50 g	
Whole black pepper	25 g	0.49
Sodium nitrate	25 g	
Total ingredients	20.9 kg	100.00

Yield on finished product: 85 per cent on recipe ingredients

Estimated carcase cut-out of meat inputs: 82 per cent

Carcase meat "bone-in" required: 20 kg ÷ 0.82 = 24.39 kg

Yield on carcase meat "bone-in" : ((20.9 x 0.85) ÷ 24.39) x 100 = 73 per cent

Drying and smoking

The sausage is held at 15°C for a period of 24 hours in a drying room and then placed in a smokehouse. It is smoked for 24 hours at 25°C, after which the smokehouse temperature is raised slowly to 45°C. The product is then further smoked for six hours. It should have a golden brown colour.

Cooling and dispatch

The product is held at 25°C for four hours and transferred to a holding cold store (5°C). It is then over-packed with cellophane and dispatched, or may be held for up to five days (at the processing unit) at 5°C for maturing.

The complete production sequence of beef cervelat is illustrated in figure IV.15.

ESTIMATE MEAT REQUIREMENT   ESTIMATE CASING REQUIREMENT
            |
    PRIMAL BUTCHERY   Beef middles (or
                                              No. 2 fibrous casings)
   Neck, leg, shoulder
        / | \
Trimmings Lean meat   Fat

Weighted to give 85% lean meat,
        15% fat
         _____|_____
        | |
    Lean meat   Trimmings and fat
        | |
  CHILL, 0°C, 5 hrs   CHILL 0°C, 4 hrs   SOAK 25°C, 2 hrs
        | |
   MINCE 6 mm   MINCE 8 mm
          \   /
            \ MIX /

Salt   )
Sugar   )
Ground pepper)
Sodium nitrate)

              MINCE, 3 mm
                 |
Whole black pepper —KNEAD (MIX), 2-5 mn
                 |
            CURE 5°C, 72 hrs

RE-MIX ——→ STUFF ————————————————————————
                 |
         LINK (35 cm portions)
                 |
         DRYING 15°C, 24 hrs
                 |
         SMOKE I, 25°C, 24 hrs
                 |
         SMOKE II, 45°C, 6 hrs
                 |
         COOL 25°C, 4 hrs
                 |
         COOL 5°C, 24 hrs
                 | MATURE 0°C, 5 days
         PACK (cellophane) |
                 | PACK (cellophane, vacuum)
         DISPATCH |
                                      DISPATCH

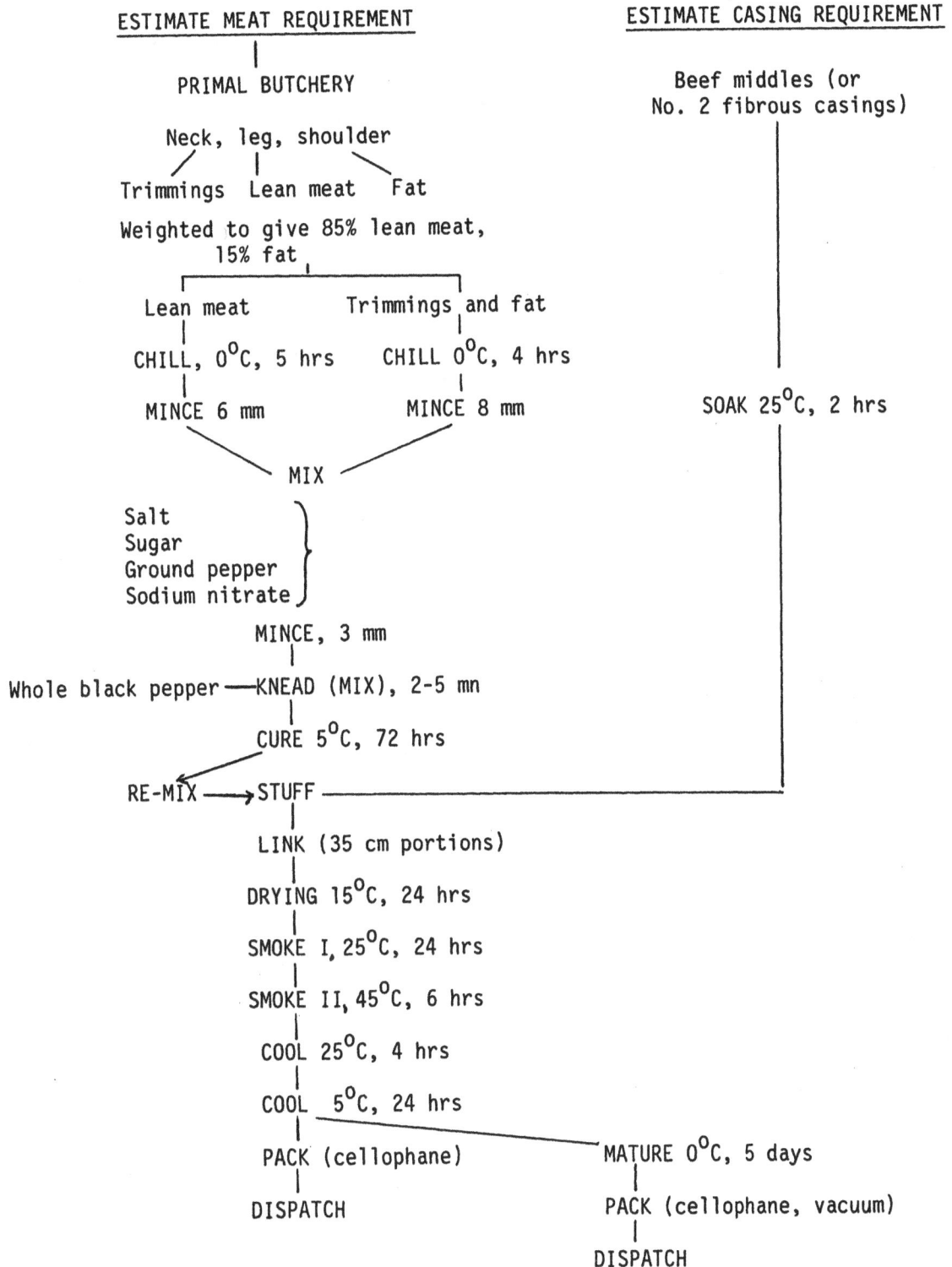

Processing time: 160 hrs

Yield on "bone-in" meat: 73 per cent

Figure IV.15

Flow diagram illustrating beef cervelat preparation

CHAPTER V

EVALUATION OF ALTERNATIVE
MEAT PROCESSING TECHNIQUES

I. INTRODUCTION

The purpose of this chapter is to describe a simple evaluation methodology which may be used by practising or would-be meat processors for the evaluation of alternative processing techniques and scales of production. Section II describes, in general terms, the methodology while section III illustrates its application to the two production models described in Chapter III. Section IV analyses additional factors which should be taken into consideration when evaluating projects in this sector.

Findings from the evaluation of the production models (section III) do not necessarily appply to all situations and the reader is not advised to make investment decisions on the basis of these findings. The following three points should be borne in mind:

- the evaluations, though based on data from developing countries, are only intended to illustrate principles and methods. They do not indicate the financial viability or non-viability of the production methods under all possible conditions;

- because the provisions and effects of these models vary from country to country and with each individual situation, no account has been taken in the evaluation of taxation, eligibility for government subsidies or special measures favouring small-scale enterprises. These factors must be taken into account where appropriate; and

- it will be necessary for potential producers to calculate for themselves the capacities required (buildings, equipment, and so on) and the types and levels of operating inputs needed, according to market size and the product mix required.

A beef processing plant requires some or all of the following elements for its effective establishment and operation:

- a suitable site (e.g. in terms of closeness to raw materials and/or markets);
- a reliable external source of electricity or the capacity to generate it;
- a reliable source of treated water;
- qualified labour for some of the operations;
- managerial and technical expertise;
- reliable sources of raw materials; and
- a market for the product.

It is important that potential entrepreneurs check all the above elements prior to deciding whether to invest in any particular plant.

The two production models evaluated in this chapter have the following characteristics:

Model 1: This plant is designed to produce 50 tonnes of beef products per annum, including 25 tonnes of beefburgers, 18.75 tonnes of chili con carne and 6.25 tonnes of charqui. The plant employs six workers in addition to the manager and operates 250 working days per year (50 weeks), for eight hours per day (one shift). Average weekly production is therefore equal to 1 tonne of products;

Model 2: This plant is designed to produce the same mix of products as that of Model 1 but operates two shifts per day instead of one. The total yearly production is thus 100 tonnes, including 50 tonnes of beefburgers, 37.5 tonnes of chili con carne and 12.50 tonnes of charqui.

In both models the following assumptions are made:
- that the plant is run as an entirely separate operation from the slaughterhouse or butchery business;

- that supplies of whole beef carcases are delivered for processing from a nearby slaughterhouse (all inputs are therefore of "bone-in" meat); and

- that a suitable system is available for distributing and retailing the finished beef products at additional costs (i.e. revenues are based on factory-gate prices).

The evaluation is therefore concerned with production from the point where carcases are delivered to the plant to the point where the finished beef products leave the factory.

II. EVALUATION METHODOLOGY

The staff of financial institutions, businessmen and government officials may have their own evaluation methodology but may still find the one described in this memorandum useful, especially if they are unfamiliar with the processing of beef.

The evaluation framework consists of three main parts:

- the determination of quantities of various inputs used in the processing of beef (steps 1 to 5);
- the estimation of the cost of each input and that of unit production costs (steps 6 to 13); and
- the calculation of the project profitability (steps 14 and 15).

These steps are described below. Producers wishing to identify the most appropriate technique should repeat these steps for each technology which may yield the required output.

Step 1: Determination of the quantity and type of processed meat to be produced each year. These are a function of market demand, availability of investment funds, adopted production techniques, availability of raw materials and so on.

Step 2: Estimation of the quantities of the various material inputs for the adopted scale of production and product mix. The main materials for beef processing are:

- beef (various cuts, trimmings);
- salt;
- sugar;
- nitrite and nitrate;
- spices and herbs;
- casings;
- water;
- packaging materials; and
- power (eletricity).

Step 3: Compilation of a list of required equipment, including spare parts and servicing and testing equipment. Both locally made and imported equipment should be included.

Step 4: Determination of labour requirements. It should be noted that the productivity of the labour force may be significantly different from one region to another. Labour requirements are dependent on the number of shifts per day, working days per week and working weeks per year. In addition, the number of workers should be specified for each skill level.

Step 5: Determination of the local infrastructure required. This may include:

- land for the plant; and
- buildings, including processing hall, storage areas, office and so on.

Step 6: Determination of the working capital required since operating costs would be incurred before revenues from sales are realised. In general, working capital is estimated at 25 per cent of annual operating costs (cost of materials plus that of labour).

Step 7: Determination of the annual depreciation costs of equipment and buildings. Whatever the type of equipment used, it will have a limited life (five to 10 years). An estimate must thus be made of the annual equipment depreciation costs. The same applies to buildings, which may be taken to have a life of 25 years. Depreciation costs are dependent on initial purchase price, the life of equipment and buildings and the prevailing interest rate. Table V.1, located at the end of this chapter, may be used to estimate these costs. It provides the discount factor (F) for interest rates up to 40 per cent and expected life periods up to 25 years. For example, for an interest rate of 10 per cent and an expected life of five years, the discount factor F is equal to 3.791. If Z is the purchase price of the equipment or cost of buildings, the annual depreciation cost is equal to Z/F. Hence, the longer the useful life, the lower the annual depreciation cost, and the higher the prevailing interest rate, the higher the depreciation cost.

The c.i.f. prices of imported pieces of equipment may be obtained from local importers or equipment suppliers. The prices of local equipment and buildings may be obtained from local contractors and equipment manufacturers.

Step 8: Determination of the annual cost of land: land has an infinite life. Thus, the annual cost of land may be assumed to be equal to the annual rental cost of equivalent land.

Step 9: Calculation of the annual cost of materials identified in step 2.

Step 10: Calculation of the annual cost of labour. This cost varies greatly from one region to another and must be calculated on the basis of local wage rates. The labour requirements are those identified in step 4.

Step 11: Working capital raised for the project will require an allowance in the annual costs for interest payments to be made on this capital.

Step 12: The total annual production cost consists of the sum of the separate annual costs itemised in steps 7 to 11.

Total annual production cost = Depreciation cost of equipment and buildings + annual land rental cost + annual labour costs + annual cost of materials + annual interest payments on working capital.

Step 13: The average unit production cost for the adopted product mix is equal to:

Total annual cost ÷ annual output

The unit production cost of any individual meat product may be estimated as follows:

(i)   calculate the sum of cost items estimated in steps 7, 8, 10 and 11;

(ii)   divide the above sum by the total number of tonnes of meat processed per year; this gives the unit production cost, excluding unit materials cost (from step 9);

(iii) estimate the total cost of materials (meat, spices, etc.) used in the manufacture of the meat product;

(iv)   estimate the unit materials cost by dividing the total materials costs by the number of tonnes produced each year; and

(v)   add the unit materials cost to the unit cost estimated under (ii) above.

Step 14: Estimation of total revenues from sales on the basis of an estimate of wholesale or retail prices of the various meat products produced by the plant.

Step 15: Estimation of total profits before taxes. These are equal to total revenues minus total annual production costs.

III. APPLICATION OF THE EVALUATION METHODOLOGY

The evaluation methodology is applied to the two production models specified in the previous section.

III.1 Model 1: 50 tonnes per year beef processing plant

Step 1:

   Plant capacity : 50 tonnes/year

   Product mix: - 25 tonnes of beefburgers

              - 18.75 tonnes of chili con carne

              - 6.25 tonnes of charqui

Step 2: Yearly material inputs (estimated on the basis of data from Chapters III and IV):

- carcase beef ("bone-in")	68.44 tonnes
- salt, nitrite salt	2.87 tonnes
- spices, herbs	1.453 tonnes
- cassava flour	1.32 tonnes
- onion	823 kg
- packaging materials	18.75 tonnes for <u>chili con carne</u>
	31.25 tonnes for other products
- electricity	55,100 kW
- water	1.5 million litres

<u>Step 3</u>: List of equipment required, including life of each piece of equipment : see Appendix III.

<u>Step 4</u>: Labour requirements

The plant operates one eight-hour shift per working day for a total of 250 working days per year. The personnel required include the following:

- 1 manager;
- 2 butchers;
- 2 processors;
- 1 processing assistant; and
- 1 unskilled worker;

Total number of staff: 7

<u>Step 5</u>: Land and building requirements

Total land required : 2,000 m^2

Buildings required (see Chapter III for layout):

Item	Area (m^2)
Processing and butchery area	160
Drying room	20
Packaging room, dispatch, stores (including cold store area)	93
Maintenance and boiler room (separate)	15
Office, toilets	15
Salting area and store	40
Cold storage modules	5
Concrete apron, hard standing and access	

<u>Step 6</u>: Working capital required is equal to three months operating costs (for labour, materials and maintenance). This amounts to 25 per cent of the sum calculated in steps 7, 9 and 10 below:

(US$6,092 + US$120,630 + US$20,000) = <u>$36,680</u>

Step 7: Annual depreciation costs of equipment and buildings

- Equipment: The interest rate used for the calculation of equipment depreciation costs is assumed to be 12 per cent. The lives of the various pieces of equipment are given in Appendix III. The total cost of equipment with lives of 5, 10 and 15 yers respectively is as follows:

Cost of equipment (US$)	Useful life (years)	Discount factor (F)	Depreciation costs (US$)
18 700	5	3.605	5 187
21 100	10	5.650	3 735
23 450	15	6.811	3 443
63,250			

Total annual depreciation cost of equipment: 12 365

- Buildings: The total cost of buildings is estimated at US$58,600[1]. Their assumed useful life is 25 years. Thus, for a 12 per cent interest rate, the discount factor (F) is equal to 7.843. The buildings depreciation cost is then equal to:

$$US\$ \ 58,600 \div 7.843 \ = \ US\$ \ 7,472$$

- Total annual depreciation costs: To the annual equipment and buildings depreciation costs should be added the annual cost of spares and maintenance. The latter is estimated at 5 per cent of the cost of equipment and buildings:

$$(63,250 + 58,600) \times 0.05 \ = \ US\$6,092$$

Thus, the total annual depreciation cost is equal to:

$$US\$12,365 \ + \ US\$7,472 + US\$6,092 \ = \ \$25,929$$

---

[1] This estimate is based on the following unit costs:
- US$150 per m^2 for processing and butchery area, packaging room and maintenance and boiler rooms;
- US$160 per m^2 for drying room, office and toilets;
- US$800 per m^2 for cold storage modules;
- US$ 80 per m^2 for salting area and store; and
- US$5,600 for concrete apron, hard standing and access.

Step 8: Annual rental cost of land. This is assumed to be equivalent to an interest rate of 12 per cent paid on the value of the land:
- assumed land value: US$1,000
- annual rental cost = US$1,000 x 0.12 = US$120

Step 9: Annual cost of material inputs and other miscellaneous items.

Item	Unit cost (US$)	Quantity	Total cost (US$)
Carcase meat (bone-in)	1,400/tonne	68.44 tonnes	95 820
Salt, nitrite salt	130/tonne	2.87 tonnes	370
Spices, herbs	2,150/tonne	1,453 kg	3 120
Cassava flour	100/tonne	1,320 kg	130
Onion	250/tonne	823 kg	210
Packaging[1]	0.15/kg (chili)	18.75 tonnes	2 810
	0.05/kg other	31.25 tonnes	1 560
Electricity[2]	0.10/kW	55,010 kW	5 510
Water[1]	1.30/1,000 litres	1.5 million litres	1 950
Insurance[3]			1 220
Knives[1]			430
Office expenses[1]			2 000
Detergents and cleaning utensils[1]			1 500
Sundries and unforeseen[1]			4 000
Total annual cost			120 630

Step 10: Annual labour costs

Category	Number	Cost (US$)
Manager	1	4 400
Butcher	2	6 600
Processor	2	5 500
Assistant processor	1	2 000
Labourer	1	1 500
Total	7	20 000

[1] Estimated.
[2] Estimated on the basis of data from Appendix III.
[3] Equal to 1 per cent of the value of buildings and equipment.

Step 11: Annual cost of working capital at a 12 per cent interest rate:

US$36,680 x 0.12 = US$4 401

Step 12: Total annual production cost (sum of totals from steps 7, 8, 9, 10 and 11):

US$25,929 + US$120 + US$120,630 + US$20,000 + US$4,401 = US$171 080

Step 13: Production cost per average tonne of product:

US$171,080 ÷ 50 tonnes = US$3,422/tonne

Step 14: Annual revenue from sale of output:

Product	Quantity (tonnes)	Price, ex-plant per ton. (US$)	Annual revenue (US$)
Beefburgers	25.00	3 200	80 000
Chili con carne	18.75	3 500	65 630
Charqui	6.25	4 800	30 000
Total	50.00		175 630

Step 15: Annual profits before taxes:

US$175,630 - US$171,080 = US$4,550

III.2 Model 2: 100 tonnes per year meat processing plant

Step 1: The plant has the capacity to produce 100 tonnes of beef products per annum with the following product mix:
- 50 tonnes of beefburgers;
- 37.5 tonnes of chili con carne; and
- 12.5 tonnes of charqui.

Step 2: Yearly material inputs (estimated on the basis of data from Chapters III and IV).

- carcase beef ("bone-in")   136.88 tonnes
- salt, nitrite salt   5.78 tonnes
- spices, herbs   2,906 kg
- cassava flour   2,640 kg
- onion   1,646 kg
- packaging materials   37.5 tonnes for chili con carne
                                     62.5 tonnes for other products
- electricity   69,220 kW
- water   2.7 million litres

Step 3: List of equipment required: see Appendix III.

Step 4: Labour requirements

The plant operates two shifts per working day for a total of 250 working days per annum. The personnel required include the following:

- manager   1
- deputy manager   1
- butcher   4
- processor   4
- processing assistant   2
- labourer   2
  Total   14

Step 5: Land and building requirements

Total land required: 2,000 m^2

Buildings required (see Chapter III for layout):

Item	Area (m^2)
Processing and butchery area	160
Drying room	20
Packaging room, dispatch, stores	93
Office, toilets	15
Maintenance and boiler room (separate)	15
Salting area and store	40
Cold storage modules	10
Concrete apron, hard standing and access	

Step 6: Working capital required is equal to three months operating costs (for labour, materials and maintenance). This amounts to 25 per cent of the sum of cost items estimated in steps 7, 9 and 10 below:

($12,780 + $232,410 + $39,120) x 0.25 = $71,080

Step 7: Annual depreciation costs of equipment and buildings
- Equipment: The interest rate used for the calculation of equipment depreciation costs is assumed to be 12 per cent. The lives of the various pieces of equipment are given in Appendix III. The total cost of equipment with lives of 5, 10 and 15 years respectively is as follows:

Cost of equipment (US$)	Useful life (years)	Discount factor (F)	Depreciation cost (US$)
20 650	5	3.605	5 730
21 100	10	5.650	3 735
23 450	15	6.811	3 440
65 200			

Total annual depreciation cost of equipment: 12 905

- Buildings:

Assumed useful life = 25 years. Interest rate = 12 per cent.

The total cost of buildings is assumed to be $62,600[1]. The discount factor (F) is equal to 7.843. The buildings annual depreciation cost is then equal to:

$$US\$62,600 \div 7.843 = US\$7,980$$

- Total annual depreciation costs: To the annual equipment and buildings depreciation costs should be added the annual cost of spares and maintenance. The latter is estimated at 10 per cent of the cost of equipment and buildings[2]:

$$(US\$65,200 + US\$62,600) \times 0.1 = US\$12,780$$

Thus, the total annual depreciation cost is equal to:

$$US\$12,905 + US\$7,980 + US\$12,780 = US\$33,665$$

Step 8: Annual rental cost of land. This is assumed to be equivalent to an interest rate of 12 per cent paid on the value of the land:

- assumed land value: US$1,000
- annual rental cost: 1,000 x 0.12 = US$120

Step 9: Annual cost of material inputs and other miscellaneous items:

Item	Unit cost (US$)	Quantity	Total cost(US$)
Carcase beef	1,400/tonne	136.88 tonnes	191 630
Salt, nitrite salt	130/tonne	5.78 tonnes	750
Spices, herbs	2,150/tonne	2,906 kg	6 250
Cassava flour	100/tonne	2,640 kg	260
Onion	250/tonne	1,646 kg	410

---

[1] The total cost of buildings is estimated on the basis of the unit costs used for Model 1 (see note 1 p. 88).
[2] This percentage is double that used for Model 1 in view of the higher wear and tear associated with a double shift operation.

Packaging[1]	0.15/kg (chili)	37.5 tonnes	5 620
	0.05/kg (other)	62.5 tonnes	3 120
Electricity[1]	0.1/kW	69,220 kW	6 920
Water	1.3/1,000 litres	2.7 million litres	3 510
Insurance[2]			1 280
Knives[3]			860
Office expenses[3]			3 000
Detergents, cleaning utensils[3]			2 800
Sundries and unforeseen[3]			6 000
- Total annual cost			232 410

Step 10 : Annual labour costs:

Category	Number	Cost (US$)
Manager	1	4 400
Deputy manager	1	3 520
Butcher	4	13 200
Processor	4	11 000
Processing assistant	2	4 000
Labourer	2	3 000
Total	14	39 120

Step 11: Annual cost of working capital at 12 per cent interest rate:

US$71,080 x 0.12 = US$8,530

Step 12: Total annual production cost (sum of totals from steps 7, 8, 9, 10 and 11):

US$33,665 + US$120 + US$232,410 + US$39,120 + US$8,530 = US$313,845

Step 13: Production cost per average tonne of product:

US$313,845 ÷ 100 tonnes = US$3,138/tonne

---

[1] Estimated on the basis of data from Appendix III.

[2] Equal to 1 per cent of the value of buildings and equipment.

[3] Estimated.

Step 14: Annual revenues from sale of output:

Product	Price, ex-plant per tonne (US$)	Quantity (tonnes)	Annual revenue (US$)
Beefburger	3 200	50.00	160 000
Chili con carne	3 500	37.50	131 250
Charqui	4 800	12.50	60 000
Total		100.00	351 250

Step 15: Annual profits before taxes:

$$US\$351,250 - US\$313,845 = \underline{US\$37,405}$$

IV. OTHER CONSIDERATIONS IN PROJECT EVALUATION

IV.1 Sensitivity analysis

Any project evaluation makes use of various assumptions regarding the prices of raw materials, the retail prices of the output, the interest rate and so on. It is not certain that these assumptions will actually prevail after the start of the project. It is therefore advisable to repeat the evaluation with alternative sets of "pessimistic" assumptions (e.g. higher prices of raw materials and equipment, higher interest rates, lower retail prices of the output). The entrepreneur may then have an idea of the risk he will be taking by investing in a given project. It is, in particular, important to repeat the evaluation for various beef prices since the cost of meat inputs is by far the most important cost item. Entrepreneurs will then be able to know the maximum price they may afford to pay for raw meat while still making a profit.

IV.2 Choice of product and product price

The profitability of a meat processing plant will also depend on the choice of product mix and prevailing retail prices, the latter being a function of demand and supply conditions. For example, if markets for several alternative product mixes are available in a country, it would probably be worthwhile analysing several production models in order to identify the most profitable one.

Before returns from sales of beef products can be estimated, wholesale (ex-factory) prices must be known. In some countries, imported or domestically produced beef products may already be on the market. Retail prices will then be available for these, although they may have to be adjusted if differences in quality are being considered or if the supply of meat products is to be increased substantially. In any case, ex-factory prices could be estimated from these retail prices by deducting retail and distribution margins ("mark-ups").

If it is difficult to carry out a market survey, or if few beef products are being sold in the country, it would be prudent to set prices on the basis of estimated unit production costs and an acceptable return on investment. It should be borne in mind that steps might have to be taken later, once manufacture has begun, to correct any resulting shortages or surpluses. For example, if shortages of a particular product developed, it might be possible to increase production or raise prices. On the other hand, surpluses might be exported or production reduced and other beef products manufactured instead. Ultimately, prices might have to be reduced in order to avoid higher losses.

IV.3 Siting of the plant

The location of the plant site should affect a project's profitability in terms of both fixed and variable costs. There are basically two options: to site the plant near the slaughterhouse or near the main area of consumption, assuming that these are some distance apart. The handling and transport costs of raw meat will be lower for the first option, but distribution costs for the products will be higher. The opposite conditions will apply if the plant is to be located near the main area of consumption. Furthermore, the price of land may differ from one area to another. Entrepreneurs should therefore carefully assess the effects on both investment costs and operating costs of alternative plant sites, with a view to identifying the least-cost option.

IV.4 Integration of manufacture of beef products
with other operations

As already indicated in Chapter I, a beef processing plant may be operated as a separate enterprise or may be attached to a slaughterhouse or a butchery business. In the latter two cases, the evaluation of the meat processing project may be carried out through what is called a "marginal" analysis. In this case, additional fixed and variable costs are compared to additional revenues, taking into consideration available facilities for the slaughterhouse or butchery and current sales from these businesses.

There may be certain advantages to be gained by integrating a beef processing plant into an existing slaughterhouse complex or butchery business rather than building a separate plant. For example, there might already be sufficient land available for a plant site at no additional cost, or existing buildings might be adapted at a relatively low cost. Transport facilities could also be used more effectively in the servicing of the slaughterhouse and processing plant.

A small plant attached to a butchery business may also improve the profitability of the latter. Excess supplies of fresh meat may be processed instead of being kept frozen or sold at low prices. Hired labour may be used in the meat processing unit during idle times. The investment costs will, in this case, be limited to the acquisition of a few pieces of equipment and the expansion of existing buildings.

There are also various advantages derived from the setting-up of a separate beef processing plant, including those derived from economies of scale. Thus, would-be meat processors are advised to analyse carefully the advantages and disadvantages of alternative options if funding is available for all of these. The evaluation framework described in this chapter could be used for the identification of the most profitable option.

Table V.1

Discount factor (F)

| Year | \multicolumn{18}{c}{Interest rate (percentage)} ||||||||||||||||||
	5	6	8	10	12	14	15	16	18	20	22	24	25	26	28	30	35	40
1	0.952	0.943	0.926	0.909	0.893	0.877	0.870	0.862	0.847	0.833	0.820	0.806	0.800	0.794	0.781	0.769	0.741	0.714
2	1.859	1.833	1.783	1.736	1.690	1.647	1.626	1.605	1.566	1.528	1.492	1.457	1.440	1.424	1.392	1.361	1.289	1.224
3	2.723	2.673	2.577	2.487	2.402	2.322	2.283	2.246	2.174	2.106	2.042	1.981	1.952	1.923	1.868	1.816	1.696	1.589
4	3.546	3.465	3.312	3.170	3.037	2.914	2.855	2.798	2.690	2.589	2.494	2.404	2.362	2.320	2.241	2.166	1.997	1.849
5	4.330	4.212	3.993	3.791	3.605	3.433	3.352	3.274	3.127	2.991	2.864	2.745	2.689	2.635	2.532	2.436	2.220	2.035
6	5.076	4.917	4.623	4.355	4.111	3.889	3.784	3.685	3.498	3.326	3.167	3.020	2.951	2.885	2.759	2.643	2.385	2.168
7	5.786	5.582	5.206	4.868	4.564	4.288	4.160	4.039	3.812	3.605	3.416	3.242	3.161	3.083	2.937	2.802	2.508	2.263
8	6.463	6.210	5.747	5.335	4.968	4.639	4.487	4.344	4.078	3.837	3.619	3.421	3.329	3.241	3.076	2.925	2.598	2.331
9	7.108	6.802	6.247	5.759	5.328	4.946	4.772	4.607	4.303	4.031	3.786	3.566	3.463	3.366	3.184	3.019	2.665	2.379
10	7.722	7.360	6.710	6.145	5.650	5.216	5.019	4.833	4.494	4.192	3.923	3.682	3.571	3.465	3.269	3.092	2.715	2.414
11	8.306	7.887	7.139	6.495	5.938	5.453	5.234	5.029	4.656	4.327	4.035	3.776	3.656	3.544	3.335	3.147	2.752	2.438
12	8.863	8.384	7.536	6.814	6.194	5.660	5.421	5.197	4.793	4.439	4.127	3.851	3.725	3.606	3.387	3.190	2.779	2.456
13	9.394	8.853	7.904	7.103	6.424	5.842	5.583	5.342	4.910	4.533	4.203	3.912	3.780	3.656	3.427	3.223	2.799	2.468
14	9.899	9.295	8.244	7.367	6.628	6.002	5.724	5.468	5.008	4.611	4.265	3.962	3.824	3.695	3.459	3.249	2.814	2.477
15	10.380	9.712	8.559	7.606	6.811	6.142	5.847	5.575	5.092	4.675	4.315	4.001	3.859	3.726	3.483	3.268	2.825	2.484
16	10.838	10.106	8.851	7.824	6.974	6.265	5.954	5.669	5.162	4.730	4.357	4.033	3.887	3.751	3.503	3.283	2.834	2.489
17	11.274	10.477	9.122	8.022	7.120	6.373	6.047	5.749	5.222	4.775	4.391	4.059	3.910	3.771	3.518	3.295	2.840	2.492
18	11.690	10.828	9.372	8.201	7.250	6.467	6.128	5.818	5.273	4.812	4.419	4.080	3.928	3.786	3.529	3.304	2.844	2.494
19	12.085	11.158	9.604	8.365	7.366	6.550	6.198	5.877	5.316	4.844	4.442	4.097	3.942	3.799	3.539	3.311	2.848	2.496
20	12.462	11.470	9.818	8.514	7.469	6.623	6.259	5.929	5.353	4.870	4.460	4.110	3.954	3.808	3.546	3.316	2.850	2.497
21	12.821	11.764	10.017	8.649	7.562	6.687	6.312	5.973	5.384	4.891	4.476	4.121	3.963	3.816	3.551	3.320	2.852	2.498
22	13.163	12.042	10.201	8.772	7.645	6.743	6.359	6.011	5.410	4.909	4.488	4.130	3.970	3.822	3.556	3.323	2.853	2.498
23	13.489	12.303	10.371	8.883	7.718	6.792	6.399	6.044	5.432	4.925	4.499	4.137	3.976	3.827	3.559	3.325	2.854	2.499
24	13.799	12.550	10.529	8.985	7.784	6.835	6.434	6.073	5.451	4.937	4.507	4.143	3.981	3.831	3.562	3.327	2.855	2.499
25	14.094	12.783	10.675	9.077	7.843	6.873	6.464	6.097	5.467	4.948	4.514	4.147	3.985	3.834	3.564	3.329	2.856	2.499

CHAPTER VI

SOCIO-ECONOMIC AND ENVIRONMENTAL CONSIDERATIONS

The previous chapter was mostly of interest to established or potential meat processors since it dealt with the private profitability of beef processing units. This chapter reviews issues of interest to public planners and project evaluators from industrial development agencies, such as the employment effects of alternative meat processing technologies, foreign exchange expenditures, industrial location and the protection of the environment.

I. PROCESSED BEEF AND BASIC NEEDS

It is argued that the production of grain (e.g. corn) for the feeding of animals may not be justified in countries with a food shortage, since the nutritional value of a given quantity of grain is much higher than that of the equivalent amount of meat it helps produce. Thus, meat production in developing countries may be justified under three main conditions: whenever there is a surplus of agricultural products; when pasture land is available which is not fit or required for the growing of foodgrain; and when agricultural wastes are available. In many developing countries, one or more of the above conditions apply and meat is not, therefore, produced at the expense of the nutritional needs of low-income groups. This is not true for other developing countries where the raising of cattle uses large amounts of food grain which could have been made available for human consumption.

Whenever the production of fresh meat is justified from a socio-economic point of view, meat processing will also be justified for the following reasons. Firstly, a demand for various varieties of processed beef will always exist in a country. This demand may be either satisfied through imports or through local production. Obviously, the latter solution should be preferred since scarce foreign exchange will not be used for the import of processed beef. Furthermore, low-income groups will not generally be able to afford expensive imports. Secondly, the production and/or marketing of fresh meat may not be possible in some areas of a country if transport in refrigerated trucks or wagons is not available. Thus these areas may be supplied with processed beef products which will not spoil if kept at ambient temperatures for a few days or weeks. Finally, there are cases where beef trimmings or excess supplies of meat could be wasted if they are not processed. Altogether, the promotion of meat processing could help satisfy the basic need for food of low-income groups. There is, however, a danger that the price of fresh beef may increase if too much of the available supply is diverted to the production of high-priced beef products which cannot be afforded by low-income groups.

II. THE EMPLOYMENT EFFECT OF ALTERNATIVE MEAT PROCESSING TECHNOLOGIES

The generation of productive employment is one of the major development objectives of developing countries. Technologies which may help to achieve this objective should therefore be favoured as long as labour is used in an efficient manner. In the case of beef processing, the employment effect of alternative processing technologies may be analysed in relation to a hypothetical national production of 6,400 tonnes per annum of a variety of beef products. Production may be carried out in a number of small-scale plants (e.g. with a capacity of 50 tonnes to 100 tonnes per year) or in a large scale plant producing 6,400 tonnes per year (i.e. the whole national production).

Table V.1 provides estimates of employment generated by the small-scale plants and the large-scale plant for the selected hypothetical yearly output. This table shows that the small-scale plants generate twice as much employment as the large-scale plant. While the labour intensity of the small-scale plants is substantially higher than that of the large-scale plant, the additional employment which may be generated in an average size developing country through the establishment of small-scale meat processing plants in place of large-scale units should be relatively small in view of the limited volume of fresh meat processed each year. On the other hand, the fact that small-scale plants will not need foreign expertise for their operation should be of particular interest to developing countries.

Table VI.1

Employment generated by small-
and large-scale beef processing plants

Plant size (tonnes/year)	No. of plants required to produce 6,400 tonnes	Employees per plant	Total employment	Man-years per 1,000 tonnes
50	128	7	896	140
100	64	14	896	140
6,400	1	435	435	68

III. INVESTMENT COSTS AND FOREIGN EXCHANGE EXPENDITURES

Projects which require relatively low capital and foreign exchange expenditures should be preferred by developing countries which suffer from shortages of investment funds and foreign exchange for the financing of development projects. If the above criterion is to be used for the selection of beef processing technology, small-scale plants will be by far preferred, as indicated in table VI.2. This table shows that total investment costs per tonne of product for small-scale plants is approximately 25 to 50 per cent that needed for the large-scale plant. The foreign exchange costs per tonne of output is also much lower for small-scale plants (approximately 20 to 30 per cent of that required by a large-scale plant).

In addition to making a better use of scarce investment and foreign exchange resources, small-scale beef processing units may be established without government financial assistance or the need for foreign investments (e.g. in the form of outright ownership or joint ventures). Furthermore, small units may be established all over the country while large units must be established in the main urban areas since they usually require a fairly well-developed infrastructure. Thus, the adoption of small-scale technologies should allow a more balanced geographical development. This matter is further considered below.

Table VI.2

Investment costs and foreign exchange expenditures for small and large-scale beef processing plants

	Plant size (tonnes/year)		
	50	100	6,400
Investment cost per plant (US$)	121 850	127 800	30 240 000
Foreign exchange cost per plant (US$)[2]	63 250	65 200	24 192 000
No. of plants required to produce 6,400 tonnes	128	64	1
Total investment cost (US$)	15 596 800	8 179 200	30 240 000
Investment cost per tonne of output (US$)	2 437	1 278	4 725
Total foreign exchange cost (US$)	8 096 000	4 182 800	24 192 000
Foreign exchange cost per tonne of output (US$)	1 265	653.5	3 780

[1] Investment cost per plant includes the cost of land, buildings and equipment.

[2] Foreign exchange cost per plant is assumed to be equal to the cost of equipment since, in most developing countries, the equipment must be imported. These estimates should be adjusted for countries which may produce some of the equipment.

IV. LOCATION OF BEEF PROCESSING PLANTS

The location of any industry may be a function of the following:

- transport costs of inputs and outputs and the availability of transport facilities;
- the availability of qualified manpower;
- the availability of regular supplies of water and energy; and

- government plans regarding the development of various areas of a country.

The last factor may induce governments to invest in infrastructural works or to provide various incentives in order to attract industries in particular areas of the country.

In the case of beef processing, most of the above factors apply to a much lesser degree to small-scale units than to large-scale plants. The availability of good transport facilities is not essential if live or slaughtered cattle are available locally since the output of a small-scale plant can be marketed locally. Thus the need for good roads or refrigerated trucks is much reduced.

The equipment used in small-scale plants is fairly simple and does not need highly qualified manpower. Small plants need only a qualified butcher with some management training. On the other hand, large-scale plants need highly qualified technicians who may not be available in small towns or rural areas.

The availability of regular supplies of water and energy applies equally to small- and large-scale plants. However, the requirement of these for the latter plants is much more important and any disruption in the supply of water or electricity could adversely affect plant profitability. On the other hand, a small-scale plant may adjust to such disruptions by storing water in a small tank or by using a small generator. Furthermore, losses incurred by a small-scale unit as a result of disruptions in water or electricity supply are much less serious than in the case of a large-scale plant.

In summary, governments wishing to create industries in backward areas of the country should find it much easier to establish small-scale beef processing plants than large-scale plants in these areas. The establishment of large meat processing units in small towns or rural areas will require important infrastructural works which may not always be justified.

V. ENVIRONMENTAL EFFECTS OF ALTERNATIVE
    MEAT PROCESSING TECHNOLOGIES

In general, the environmental effects of food processing technologies is much more important in the case of large-scale than small-scale production.

Although the amount of pollutants per unit of output may be the same for all scales of production, the pollution generated by a large-scale plant in a given location could have more harmful effects than equivalent pollution generated by a number of small-scale plants located in various areas of a country.

However, the above general statement does not always apply. For example, a large-scale plant located in an urban area may be forced to implement various measures in order to minimise or eliminate the amount of pollution generated, since the latter may not be tolerated by the urban population. On the other hand, a small-scale plant located in a rural area or small town may not be forced to apply similar measures because the amount of pollutants is tolerable.

The various effects of meat processing technologies on the environment are reviewed below, and remedial measures for the elimination or minimisation of these effects are described.

V.1 Waste disposal

The type and quantity of wastes resulting from meat processing will depend on whether the processing unit is operated as a separate business or is attached to a slaughterhouse. In the former case, the main waste materials include bones, some inedible trimmings and small amounts of grease and blood which may be contained in water used for cleaning the equipment, work-benches, and so on. In the latter case, other waste materials must be disposed of, including blood, the content of edible offal (these are usually processed in a separate tripery), inedible offal and so on. In both cases, measures must be taken in order to dispose of these wastes in the most hygienic manner and to avoid the contamination of water sources or the spreading of disease. The disposal of the most important waste materials is briefly described below.

Blood

Blood coagulates into a solid mass soon after leaving the body of the animal. It should not, therefore, be diverted into a sewage unit unless the latter carries a sufficient volume of water to dilute the blood. The same applies to septic tanks. If blood from a slaughterhouse cannot be diverted to a main sewage system, and if no facilities are available to transform it into, for example, stock feed, it should be removed by special drains which have no connection with the sewer system. Blood may be disposed of by collecting it outside the slaughterhouse into a covered pit and letting it seep into the ground. The pit should be covered in order to avoid smells and insects. The special drain should be regularly flushed with water in order to avoid clogging of the system.

If the number of slaughtered animals is sufficiently large, it may be profitable to install a blood-processing unit for the production of stock feed. In some countries, blood may also be processed for human consumption.

Bones

Meat processing plants generate large amounts of bones which must be disposed of. In many countries, bones are transformed into animal feed components or other industrial products. Bones may then be stored in sealed plastic bags outside the meat processing plant for periodic delivery to the bone processing plant.

Burning of wastes

Inedible offals and the content of edible ones (e.g. the stomach) should be collected into bins and disposed of by any acceptable method which does not harm the environment. In the case of small slaughterhouses, solid wastes may be burned in a simple incinerator made from an oil drum. Another alternative could be to dig a simple trench, fill it with branches and burn the wastes disposed over the branches. In general, the burying of solid wastes should be avoided since it is very likely that these will be dug up and eaten by scavengers. Burying may be used if holes are sufficiently deep and wastes are covered with lime. This method should also be used for the whole carcase of infected animals which are not fit for human consumption.

V.2 Final disposal of effluents

The final disposal of wastes in septic tanks, chemical precipitation or other costly methods can be avoided by proper screening of the effluents and by the removal of grease and particles suspended in the form of sludge (e.g. by use of a grease trap made of screens of galvanised wire over a gulley trap into which the drain discharges from the slaughterhouse). The final choice of effluent disposal must rest with the local authorities. There are, however, some basic principles which should be applied in the planning and design of slaughterhouses.

Effluents which are free from fat and other solid particles may be drained into shallow beds for seeping and evaporation. This method is not, however, recommended in warm climates because of the risk of mosquito breeding and the development of bad smells. Thus, evaporation beds should be used only in exceptional circumstances. Another simple method of sewage disposal consists of the use of one or a series of soakage pits. These should be 6 metres deep, with a diameter of 2 metres, and should be covered in order to avoid bad smells. They work satisfactorily if used for small amounts of properly pre-treated effluents and if the type of soil is suitable. A third, cheap alternative is to divert effluents into long trenches (60 cm wide and 1.5 m deep) filled with large stones to a depth of at least 60 cm from the top. It is advisable to lead the effluents first into a distribution box from which they can be diverted into lateral trenches for subsoil irrigation.

Whatever the system used, care should be taken to avoid contamination of water resources (e.g. aquifers, lakes) or to create new nuisances such as bad smells or the breeding of mosquitoes.

APPENDICES

APPENDIX I

Permitted levels of selected food additives

Additive	Purpose	Maximum permitted level
Agar	Thickener	Limited by good manufacting practice (GMP)
BHA (Butylated hydroxyanisole)	Anti-oxidant	0.03 per cent of total weight, 0.01 per cent of fat content
L-ascorbic acid, iso-ascorbic acid and sodium salts	Colour enhancer	500 mg/kg (expressed as ascorbic acid)
Natural flavourings as defined in the Codex Alimentarius	Flavour enhancer	Limited by GMP
Natural smoke solutions	Flavour enhancer	Limited by GMP
Sodium citrate	Colour enhancer	Limited by GMP
Disodium 5-guanylate	Flavour enhancer	500 mg/kg (expressed as guanylic acid)
Disodium 5-inosinate	Flavour enhancer	500 mg/kg (expressed as inosinic acid)
Monosodium glutamate	Flavour enhancer	2,000 mg/kg (expressed as glutamic acid) 5,000 mg/kg /(expressed as glutamic acid)
Sodium nitrate potassium Nitrate	Preservative	500 mg/kg (expressed as sodium nitrate)
Sodium nitrite potassium nitrite	To fix colour	125 mg/kg (expressed as sodium nitrite)
Sodium and potassium phosphates	Binders	3,000 mg/kg (expressed as P_2O_5)
Glucono-delta-lactone	To accelerate colour fixing	3,000 mg/kg
Edible gelatine	Thickener	Limited by GMP

APPENDIX II

Factors used for the estimation of equipment
and labour requirements

These factors are included for information only and refer to Models 1 and 2. They are estimations of average production rates.

Operation	Rate
Primal butchery rate	330 kg/hour
Deboning[1]	
Debone to hand mincer	45 kg/hour
Debone to power mincer (100 kg/hr)	66 kg/hour
Debone to power mincer(>200 kg/hr)	75 kg/hour
Mincing	
Hand mincing	50 kg/hour
Power mincing	100 kg/hour
High power mincing	>200 kg/hour
(rate is function of capacity)	
Mixing	
Hand mixing of 2.5 kg batches	5 minutes
Power mixing of 5.0 kg batches	8 minutes
Power mixing of 10.0 kg batches	12 minutes
Power mixing of 20.0 kg batches	15 minutes
Stuffing	
Hand stuffing	60 kg/hour
Hydraulic stuffing	150 kg/hour
Linking	
Hand linking	60 kg/hour
Forming	
Beefburgers hand press	35 kg/hour
Packing	
Simple	40 kg/hour
Assisted table	80 kg/hour

[1] Expressed as boneless meat equivalents.

APPENDIX III

EQUIPMENT REQUIREMENTS FOR MODEL 1 AND MODEL 2

Item	Life (years)	Number	Capacity per unit	Rating[1] (kW per unit)	Use in eight-hour-shift[1] (per cent)
Butchery					
processing area					
Meat cutting table	10	3			
Preparation table	10	1			
Platform scales	15	1	200 kg		
Mincer	10	1	80 kg per hour	0.55	40
Mixer	10	1	30 litres[2]	0.56[2]	22
Ice making plant	10	1	50 kg per day	0.5	225
Balance	15	1	1,000 g	0.01	10
Hamburger press	5	3			
Sink unit	10	1			
Immersion heater	10	1		1.5	100
Whetstone knife sharpener	5	1		0.1	10
Processing hall					
Deep freeze cabinet	10	1		1.0	225
Preparation table	10	1			
Hand mincer	5	1			
Cooking kettle	15	1	25 litres	10.0	40
Scales 0-2000 g	15	3	0-2,000 g		
Misc. stainless steel implements	5				
Charqui area					
Flat bed balance	15	1	250 kg		
Weighing scale	15	1	0.15 kg		
Miscellaneous plastic trays	5				
Dry goods store					
Weighing scale	15	1	3,000 g		
Stainless steel table	10	1			
Assorted boxes	10				
Refrigeration units					
Fresh meat room (0°C)	5	1	8 m^3	3.0	225
Finished products store (0°C)	5	1	6 m^3	2.0	225
Packaging room					
Meat wrapping table	10	1			
Miscellaneous					
High pressure washer	15	1		2.2	20
Electrocutor	10	2		0.05	300
Platform scales	15	1	250 kg		
Office furniture	15				

[1] Estimation of electricity consumption. Figures shown in the table for rating (kW) and percentage use of equipment per eight hours are used to calculate the cost of electricity requirements. Some items are used for more than eight hours per 24 hours day and the percentage figure is therefore over 100. For example, refrigeration equipment is in use 24 hours per day and actually consumes electricity for about three-quarters of this time, hence the figure of 225 per cent per eight hours shown in the table. Refrigeration equipment is assumed to be in use seven days per week.

[2] The only difference in equipment for Model 2 is an increase in the size of the refrigeration units for larger cold store modules. This increases equipment costs by $1,950. The rating of the unit in the fresh meat holding room is increased to 3.2 kW and in the finished product store to 2.25 kW.

APPENDIX IV

INFORMATION SOURCES

SELECTED DIRECTORIES

The Almanac of the Canning, Freezing, Preserving Industries
Annual
Publisher: E.E. Judge and Sons,
          P.O. Box 866
          Westminster, Maryland 21157, United States

Food Processing Catalogue: Ingredients, Equipment and Supplies
Bi-yearly
Publisher: Putnam Publishing Co.
          11 East Delaware Place,
          Chicago, Illinois 60611, United States

SELECTED CURRENT PERIODICALS

Meat Plant Magazine
Monthly
Publisher: Albert Todoroff,
          10225 Bach boulevard,
          St. Louis, Missouri 63132, United States

Meat Processing
Monthly
Publisher: Davis Publishing Co.
          645 North Michigan Avenue
          Chicago, Illinois 60611, United States

Food Technology
Monthly
Publisher: Institute of Food Technologists
          221 North Lasalle Street,
          Chicago, Illinois 60601, United States

Journal of Food Science and Technology
Quarterly
Publisher: Association of Food Technologists (India)
Central Food Technological Research Institute
Mysore 2, India

Journal of Food Technology
Quarterly
Publisher: Blackwell Scientific Publications
Osney Mead,
Oxford OX2 OEL, United Kingdom

BIBLIOGRAPHY

Ashbrook, F.G.: Butchering, processing and preservation of meat (New York, van Nostrand, 1955).

Brandly, P.J. et al. : Meat hygiene (Philadelphia, Lea and Febiger, 1966).

Carosella, M.; Johnston W.R.; Surkiewicz, F.B.: "Bacteriological survey of Frankfurters produced at establishment under federal inspection" in Journal of Food Technology (Oxford), Vol 39, No. 1, 1976.

Food and Agriculture Organisation of the United Nations: Trade Yearbook (Rome), various issues, 1967, 1970, 1973, 1976, 1979.

_____: Animal by-products: Processing and utilisation (Rome, 1978).

_____: Guide to the safe use of food additives (Rome, 1979).

_____: Food preservation: Fish, meat equipment, Economic and Social Development Series No. 5/1 (Rome, 1979).

_____: Meat handling in under-developed countries: Slaughter and preservation (Rome, 1963).

Furia, T.: Handbook of food additives, (Cleveland, Ohio, CRC Press, 2nd ed., 1975).

Gerrard, F.: Sausage and small goods production (London, Hill, 1969).

_____: Meat technology (London, Hill, 1971).

Jay, J.M.; Shelef, L.A.: "Microbial modifications in raw and processed meats and poultry at low temperatures", in Journal of Food Technology (Oxford), Vol. 32, No. 5, 1978.

Jones, H.R.: Pollution control in meat, poultry and seafood processing (Park Ridge, Noyes Data Corporation, 1974).

Karmas, E.: Fresh meat processing (Park Ridge, Noyes Data Corporation, 1970).

_____: Meat product manufacture (Park Ridge, Noyes Data Corporation, 1970).

Price, J.F.: The science of meat and meat products (San Francisco, California, W. Freeman, 1972).

Rack, G.B.; Rinsted, R.: Hygiene in food manufacturing and handling (London, Food Trade Press Ltd., 1973).

United Nations Industrial Development Organisation: Information sources on the meat processing industry (New York, United Nations, 1976).

QUESTIONNAIRE

1. Full name...

2. Address..
    ....
    ....

3. Profession (check the appropriate case)

    Established meat processor.../__/
    If yes, indicate scale of production...............................

    Government official.../__/
    If yes, specify position...

    Employee of a financial institution................................/__/
    If yes, specify position...

    University staff member../__/

    Staff member of a technology institution.........................../__/
    If yes, indicate name of institution...............................
    ....

    Staff member of a training institution............................./__/
    If yes, specify..
    ....

    Other, specify...
    ....

4. From where did you get a copy of this memorandum?
    Specify if obtained free or bought.................................
    ....

5. Did the memorandum help you achieve the following:
    (Check the appropriate case)

    Learn about meat processing techniques you were not aware of   /__/

    Estimate unit production costs for various scales
    of production/technologies   /__/

    Order equipment for local manufacture   /__/

    Improve your current production technique   /__/

    Cut down operating costs   /__/

    Improve the quality of meat products   /__/

    Decide which scale of production/technology to
    adopt for a new meat processing plant   /__/

    If a Government employee, to formulate new measures
    and policies for the meat processing industry   /__/

    If an employee of a financial institution, to assess
    a request of a loan for the establishment of a meat
    processing plant   /__/

    If a trainer in a training institution, to use the
    memorandum as   supplementary training material   /__/

    If an international expert, to better advise counter-
    parts on meat processing technologies   /__/

6. Is the memorandum detailed enough in terms of: Yes   No

    - Description of technical aspects........................____......_____

- Costing information...____.....____

- Information on socio-economic impact....................____.....____

- Bibliographical information.............................____.....____

If some of the answers are 'No', please indicate why below or on a separate sheet:

....

....

....

7. How may this memorandum be improved if a second edition is to be published?..

....

....

8. Please send this questionnaire, duly completed to:

    Technology and Employment Branch
    International Labour Office
    CH-1211 GENEVA 22 (Switzerland)

9. In case you need additional information on some of the issues covered by this memorandum, the ILO would do its best to provide the requested information.